MATH FOR DEEP LEARNING

What You Need to Know to Understand Neural Networks

by Ronald T. Kneusel

no starch press

San Francisco

Printed in the United States of America

First printing

25 24 23 22 21   1 2 3 4 5 6 7 8 9

ISBN-13: 978-1-7185-0190-4 (print)
ISBN-13: 978-1-7185-0191-1 (ebook)

Publisher: William Pollock
Production Manager: Rachel Monaghan
Production Editors: Dapinder Dosanjh and Katrina Taylor
Developmental Editor: Alex Freed
Cover Illustrator: James L. Barry
Cover and Interior Design: Octopod Studios
Technical Reviewer: David Gorodetzky
Copyeditor: Carl Quesnel
Proofreader: Emelie Battaglia

For information on book distributors or translations, please contact No Starch Press, Inc. directly:
No Starch Press, Inc.
245 8th Street, San Francisco, CA 94103
phone: 415.863.9900; fax: 415.863.9950; info@nostarch.com; www.nostarch.com

Library of Congress Control Number: 2021939724

[S]

In memory of Tom "Fitz" Fitzpatrick (1944–2013), the best math teacher I ever had. And to all the math teachers out there—they receive far too little appreciation for all of their hard work.

About the Author

Ron Kneusel has been working with machine learning in industry since 2003 and completed a PhD in machine learning from the University of Colorado, Boulder, in 2016. Ron has three other books: *Practical Deep Learning: A Python-Based Introduction* (No Starch Press), *Numbers and Computers* (Springer), and *Random Numbers and Computers* (Springer).

About the Technical Reviewer

David Gorodetzky is a research scientist who works at the intersection of remote sensing and machine learning. Since 2011 he has led a small research group within a large government-services engineering firm that develops deep learning solutions for a wide variety of problems in remote sensing. David began his career in planetary geology and geophysics, detoured into environmental consulting, then studied paleoclimate reconstruction from polar ice cores in graduate school, before settling into a career in satellite remote sensing. For more than 15 years he was a principal consultant for a software services group developing image analysis and signal processing algorithms for clients across diverse fields, including aerospace, precision agriculture, reconnaissance, biotech, and cosmetics.

BRIEF CONTENTS

CONTENTS IN DETAIL

3
MORE PROBABILITY   41

4
STATISTICS   67

8
MATRIX CALCULUS   193

9
DATA FLOW IN NEURAL NETWORKS   221

10
BACKPROPAGATION
243

11
GRADIENT DESCENT
271

APPENDIX: GOING FURTHER
305

INDEX
309

FOREWORD

Artificial intelligence (AI) is ubiquitous. You need look no further than the device in your pocket for evidence—your phone now offers facial recognition security, obeys simple voice commands, digitally blurs backgrounds in your selfies, and quietly learns your interests to give you a personalized experience. AI models are being used to analyze mountains of data to efficiently create vaccines, improve robotic manipulation, build autonomous vehicles, harness the power of quantum computing, and even adjust to your proficiency in online chess. Industry is adapting to ensure state-of-the-art AI capabilities can be integrated into its domain expertise, and academia is building curriculum that exposes concepts of artificial intelligence to each degree-based discipline. An age of machine-driven cognitive autonomy is upon us, and while we are all consumers of AI, those expressing an interest in its development need to understand what is responsible for its substantial growth over the past decade. Deep learning, a subcategory of machine learning, leverages very deep neural networks to model complicated systems that have historically posed problems for traditional, analytical methods. A newfound practical use of these deep neural networks is directly responsible for this surge in development of AI, a concept that most would attribute to Alan Turing back in the 1950s. But if deep learning is the engine for AI, what is the engine for deep learning?

Deep learning draws on many important concepts from science, technology, engineering, and math (STEM) fields. Industry recruiters continue to seek a formal definition of its constituents as they try to attract top talent with more descriptive job requisitions. Similarly, academic program coordinators are tasked with developing the curriculum that builds this skill set as it permeates across disciplines. While inherently interdisciplinary in practice, deep learning is built on a foundation of core mathematical principles from probability and statistics, linear algebra, and calculus. The degree to which an individual must embrace and understand these principles depends on the level of intimacy one expects to have with deep learning technologies.

For the implementer, *Math for Deep Learning* acts as a troubleshooting guide for the inevitable challenges encountered in deep neural network implementation. This individual is typically concerned with efficient implementation of preexisting solutions with tasks including identification and procurement of open source code, setting up a suitable work environment, running any available unit tests, and finally, retraining with relevant data for the application of interest. These deep neural networks may contain tens or hundreds of millions of learnable parameters, and assuming adequate user proficiency, successful optimization relies on sensitive hyperparameter selection and access to training data that sufficiently represents the population. The first (and second, and third) attempt at implementation often requires a daunting journey into neural network interrogation, which requires dissection into and higher-level understanding of the mathematical drivers presented here.

At some point, the implementer usually becomes the integrator. This level of expertise requires some familiarity with the desired application domain and a lower-level understanding of the building blocks that enable deep learning. In addition to the challenges faced in basic implementation, the integrator needs to be able to generalize core concepts to mold a mathematical model to the desired domain. Disaster strikes again! Perhaps the individual experiences the exploding-gradient problem. Maybe the integrator desires a more representative loss function that may pose differentiability issues. Or maybe, during training, the individual recognizes that the selected optimization strategy is ineffective for the problem. *Math for Deep Learning* fills a void within the community by offering a coherent overview of the critical mathematical concepts that compose deep learning and helps overcome these obstacles.

The integrator becomes the innovator when comfort with the subject matter allows the individual to be truly creative. With innovation comes the need for information dissemination, often requiring time away from practical development for publication, presentation, and a fair amount of teaching. *Math for Deep Learning* serves as a handbook to the foundation that the innovator holds in high esteem, providing quick references and reminders of seeds that yield new developments in artificial intelligence.

Just as these roles build upon each other, deep learning creates its own hierarchy, one of nonintuitive concepts or features that solve a specific task. The sheer scope of the problem can be overwhelming without dedicated focus. Dr. Kneusel has over 15 years of industry experience applying machine learning and deep learning to image generation and exploitation problems, and he created *Math for Deep Learning* to consolidate and emphasize what matters most: the mathematical foundation from which all neural network solutions are made possible. No textbook is complete, and this one presents other resources that expound on the topics of statistics, linear algebra, and calculus. *Math for Deep Learning* is for the individual seeking a self-contained, concentrated overview of the components that build the mathematical engine for AI's primary tool.

Derek J. Walvoord, PhD

ACKNOWLEDGMENTS

I am a Bear of Very Little Brain, and long words Bother me.
—Winnie the Pooh

This book isn't just the result of my own efforts. Sincere thanks and acknowledgment are in order.

First, I want to thank all the excellent folks at No Starch Press for the opportunity to work with them again. They are all genuinely consummate professionals and a joy to interact with—and that goes double for my editor, Alex Freed. Once again, she has taken my rambling prose and finessed it into something clear and coherent.

I also want to thank my friend David Gorodetzky for his expert technical review. David's suggestions, and subtle way of pointing out goofs, have made the book stronger. If any errors remain, they are entirely my fault for not being wise enough to listen to David's sage advice.

INTRODUCTION

Math is essential to the modern world. Deep learning is also rapidly becoming essential. From the promise of self-driving cars to medical systems detecting fractures better than all but the very best physicians, to say nothing of increasingly capable, and possibly worrisome, voice-controlled assistants, deep learning is everywhere.

This book covers the essential math for making deep learning comprehensible. It's true that you can learn the toolkits, set up the configuration files or Python code, format some data, and train a model, all without understanding *what* you're doing, let alone the math behind it. And, because of the power of deep learning, you'll often be successful. However, you won't *understand*, and you shouldn't be satisfied. To understand, you need some math. Not a lot of math, but some specific math. In particular, you'll need working knowledge of topics in probability, statistics, linear algebra, and differential calculus. Fortunately, those are the very topics this book happens to address.

Who Is This Book For?

This is not an introductory deep learning book. It will not teach you the basics of deep learning. Instead, it's meant as an adjunct to such a book. (See my book *Practical Deep Learning: A Python-Based Introduction* [No Starch Press, 2021].) I expect you to be familiar with deep learning, at least conceptually, though I'll explain things along the way.

Additionally, I expect you to bring certain knowledge to the table. I expect you to know high school mathematics, in particular algebra. I also expect you to be familiar with programming using Python, R, or a similar language. We'll be using Python 3.*x* and some of its popular toolkits, such as NumPy, SciPy, and scikit-learn.

I've attempted to keep other expectations to a minimum. After all, the point of the book is to give *you* what you need to be successful in deep learning.

About This Book

At its core, this is a math book. But instead of proofs and practice exercises, we'll use code to illustrate the concepts. Deep learning is an applied discipline that you need to do to be able to understand. Therefore, we'll use code to bridge the gap between pure mathematical knowledge and practice.

The chapters build one upon the other, with foundational chapters followed by more advanced math topics and, ultimately, deep learning algorithms that make use of everything covered in the earlier chapters. I recommend reading the book straight through and, if you wish, skipping topics you're already familiar with as you encounter them.

Chapter 1: Setting the Stage   This chapter configures our working environment and the toolkits we'll use, which are those used most often in deep learning.

Chapter 2: Probability   Probability affects almost all aspects of deep learning and is essential to understanding how neural networks learn. This chapter, the first of two on this subject, introduces fundamental topics in probability.

Chapter 3: More Probability   Probability is so important that one chapter isn't enough. This chapter continues our exploration and includes key deep learning topics, like probability distributions and Bayes' theorem.

Chapter 4: Statistics   Statistics make sense of data and are crucial for evaluating models. Statistics go hand in hand with probability, so we need to understand statistics to understand deep learning.

Chapter 5: Linear Algebra   Linear algebra is the world of vectors and matrices. Deep learning is, at its core, linear algebra–focused. Implementing neural networks is an exercise in vector and matrix mathematics, so it is essential to understand what these concepts represent and how to work with them.

Chapter 6: More Linear Algebra   This chapter continues our exploration of linear algebra, focusing on important topics concerning matrices.

Chapter 7: Differential Calculus   Perhaps the most fundamental concept behind the training of neural networks is the gradient. To understand the gradient, what it is and how to use it, we must know how to work with derivatives of functions. This chapter builds the foundation necessary to understand derivatives and gradients.

Chapter 8: Matrix Calculus   Deep learning manipulates derivatives of vectors and matrices. Therefore, in this chapter we generalize the concept of a derivative to these objects.

Chapter 9: Data Flow in Neural Networks   To understand how neural networks manipulate vectors and matrices, we need to understand how data flows through the network. That's the subject of this chapter.

Chapter 10: Backpropagation   Successful training of neural networks usually involves two algorithms that go hand in hand: backpropagation and gradient descent. In this chapter, we work through backpropagation in detail to see how the math we learned earlier in the book applies to the training of actual neural networks.

Chapter 11: Gradient Descent   Gradient descent uses the gradients that the backpropagation algorithm provides to train a neural network. This chapter explores gradient descent, beginning with 1D examples and progressing through to fully connected neural networks. It also describes and compares common variants of gradient descent.

Appendix: Going Further   We must, of necessity, gloss over many topics in probability, statistics, linear algebra, and calculus. This appendix points you toward resources that will aid you in going further with the mathematics behind deep learning.

You can download all the code from the book here: *https://github.com/ rkneusel9/MathForDeepLearning/*. And please look at *https://nostarch.com/ math-deep-learning/* for future errata. Let's get started.

1

SETTING THE STAGE

Although this book has no traditional math exercises, we do need to play around with the concepts if we want to master them. We'll have plenty of opportunities for that, but instead of pencil and paper exercises, we'll use code.

This chapter will help you set the stage by configuring our working environment. Throughout the book, I'll work in Linux, specifically Ubuntu 20.04, though what we're doing most likely will work in later versions of Ubuntu and most other Linux distributions as well. For completeness, I've included sections on configuring macOS and Windows environments. I should point out that the expected operating system for deep learning is Linux, with most things working under macOS as well. Windows is typically an afterthought, and many deep learning toolkit ports are poorly maintained, though this is improving with time.

We'll begin with some instructions for installing the expected software packages. Then we'll take a quick look at the NumPy library for Python 3.x. NumPy is foundational to virtually all scientific uses of Python, and it's essential that you know how to work with it at a basic level. Next, I'll introduce SciPy. This is also a necessary toolkit for science, but we'll only need the tiniest portion of it here. Finally, I'll talk a bit about the Scikit-Learn toolkit, abbreviated here as sklearn. This valuable toolkit implements many of the traditional machine learning models.

Throughout the book, I'll often use running examples to illustrate concepts. All code snippets assume the following line has been executed:

```
import numpy as np
```

Also, in some places, the code will reference output from a snippet that appeared earlier in the chapter. The code examples are brief, so following one to the next shouldn't be burdensome. I do recommend leaving a single Python session running as you work through a chapter, though this is not required.

Installing the Toolkits

The end goal of this section is to have the following toolkits installed with *at least* the version number listed:

- Python 3.8.5
- NumPy 1.17.4
- SciPy 1.4.1
- Matplotlib 3.1.2
- Scikit-Learn (sklearn) 0.23.2

Later versions than these almost certainly will work as well.

Let's take a quick look at how we can install each of these toolkits in the major operating systems.

Linux

For the following, the $ prompt represents the command line, whereas >>> is the Python prompt.

A fresh install of Ubuntu 20.04 desktop gives us Python 3.8.5 for free. Use the code

```
$ cat /etc/os-release
```

to verify your operating system version and use python3 to run Python, as python alone starts the older Python 2.7.

These commands install NumPy, SciPy, Matplotlib, and sklearn:

```
$ sudo apt-get install python3-pip
$ sudo apt-get install python3-numpy
$ sudo apt-get install python3-scipy
$ sudo pip3 install matplotlib
$ sudo pip3 install scikit-learn
```

Test the installation by starting Python 3 and importing each module: numpy, scipy, and sklearn. Then print the __version__ string to make sure it meets or exceeds the versions listed above. For example, see the following code.

```
>>> import numpy; numpy.__version__
'1.17.4'
>>> import scipy; scipy.__version__
'1.4.1'
>>> import matplotlib; matplotlib.__version__
'3.1.2'
>>> import sklearn; sklearn.__version__
'0.23.2'
```

macOS

To install Python 3.*x* for Macintosh, go to *https://www.python.org/*, and un-der **Downloads**, choose **Mac OS X**. Then select the latest stable Python 3 re-lease. At the time of this writing, it's 3.9.2. When the download is complete, run the installer to get Python 3.9.2 set up.

After the installation, open a terminal window and verify the installation with the following:

```
$ python3 --version
Python 3.9.2
```

Assuming Python 3 installed correctly, now we can install the libraries using the terminal window and pip3, which the installer set up for us:

```
$ pip3 install numpy --user
$ pip3 install scipy --user
$ pip3 install matplotlib --user
$ pip3 install scikit-learn --user
```

And, finally, we can check the versions of the libraries from within Python 3. Enter **python3** in the terminal to pull up a Python console, and then import numpy, scipy, matplotlib, and sklearn and print the version strings, as we did above, to verify that they meet or exceed the minimum versions.

Windows

To install Python 3 and the toolkits for Windows 10, use the following steps:

1. Go to *https://www.python.org/* and click **Downloads** and **Windows**.
2. At the bottom of the page, select the x86-64 executable installer.
3. Run the installer, choosing the default options.
4. Select **Install for All Users** and **Add Python to the Windows PATH**. This is important.

When the installer finishes, Python will be available from the command prompt because we told the installer to add Python to the PATH environment variable. Therefore, open the command prompt (WINDOWS-R, cmd), and enter python. If all goes well, you'll be greeted by the Python startup message and see a >>> interactive prompt. At the time of this writing, the version installed was 3.8.2. Note that to exit Python in Windows, use CTRL-Z, not CTRL-D.

The Python installer did us the courtesy of installing pip as well. We can use it directly from the Windows command prompt to install the libraries we need. At the prompt, enter the following lines to install the NumPy, SciPy, Matplotlib, and sklearn libraries:

```
> pip install numpy
> pip install scipy
> pip install matplotlib
> pip install sklearn
```

For me, this installed NumPy 1.18.1, SciPy 1.4.1, Matplotlib 3.2.1, and sklearn 0.22.2, which meet the minimum versions above, so all is well.

To test things, start Python from the command prompt and import numpy, scipy, matplotlib, and sklearn. All three should load without error. To write Python code, install any editor you're comfortable with, or simply use Notepad.

With your toolkits installed and good to go, let's take a quick look at each library to become at least a bit more familiar with them. We'll see examples throughout the book, but I recommend you look at the suggested documentation. It's worth it.

NumPy

We installed NumPy in the previous section. Now I'll introduce some basic NumPy concepts and manipulations. A full tutorial is available online at *https://docs.scipy.org/doc/numpy/user/quickstart.html*.

Start Python. Then try the following at the prompt:

```
>>> import numpy as np
>>> np.__version__
'1.16.2'
```

The first line loads NumPy and sets up a shortcut name for it, np. Using the shortcut name isn't required, but doing so is nearly universal. We'll assume np going forward. The second line displays the version. It should be at least what is shown above.

Defining Arrays

NumPy operates on arrays and is quite good at turning lists into arrays. Think about the sort of arrays one finds in a language like C or Java. NumPy provides an advantage because although Python is elegant, it's too slow for scientific uses when simulating arrays with lists. Actual arrays are much faster. Here's an example defining an array from a list, and then examining some of its properties:

```
>>> a = np.array([1,2,3,4])
>>> a
    array([1, 2, 3, 4])
>>> a.size
    4
>>> a.shape
    (4,)
>>> a.dtype
    dtype('int64')
```

This example defines a list of four elements and then passes it to `np.array` to turn it into a NumPy array. Basic array properties include the size and the shape. The size is four elements. The shape is also four, as a tuple, showing that a is a vector, a one-dimensional (1D) array. The shape is four because array a has four elements. If a was two-dimensional (2D), the shape would have two values, one for each axis of the array. See the following example, where the shape of b tells us that b has two rows and four columns:

```
>>> b = np.array([[1,2,3,4],[5,6,7,8]])
>>> print(b)
[[1 2 3 4]
 [5 6 7 8]]
>>> b.shape
(2, 4)
```

Data Types

Python numeric data types come in two flavors: integers of arbitrary size (try 2**1000) or floating-point numbers. NumPy, however, allows for arrays of many different types. Under the hood, NumPy is implemented in C, so it supports the same set of data types C supports. The previous example shows that the `np.array` function took the given list and, since every element of the list was an integer, created an array where each element was a signed 64-bit integer. Table 1-1 has the data types NumPy works with; we can let NumPy choose the data type for us, or we can specify it explictly.

Table 1-1: NumPy Data Type Names, C Equivalents, and Range

NumPy name	Equivalent C type	Range
float64	double	$\pm[2.225 \times 10^{-308}, 1.798 \times 10^{308}]$
float32	float	$\pm[1.175 \times 10^{-38}, 3.403 \times 10^{38}]$
int64	long long	$[-2^{63}, 2^{63}-1]$
uint64	unsigned long long	$[0, 2^{64}-1]$
int32	long	$[-2^{31}, 2^{31}-1]$
uint32	unsigned long	$[0, 2^{32}-1]$
uint8	unsigned char	$[0, 255 = 2^{8}-1]$

Let's look at some examples of arrays with specific data types:

```
>>> a = np.array([1,2,3,4], dtype="uint8")
>>> a.dtype
dtype('uint8')
>>> a = np.array([1,2,3,4], dtype="int16")
>>> a = np.array([1,2,3,4], dtype="uint32")
>>> b = np.array([1,2,3,4.0])
>>> b.dtype
dtype('float64')
>>> b = np.array([1,2,3,4.0], dtype="float32")
>>> c = np.array([111,222,333,444], dtype="uint8")
>>> c
array([111, 222,  77, 188], dtype=uint8)
```

The examples with array a use integer types, and the examples with array b use floating-point types. Notice that the first b example defaulted to a 64-bit float. NumPy did this because one of the input list elements was a float (4.0).

The last example defining array c seems to be a bug. But it isn't. NumPy doesn't warn us if the requested data type can't hold the given values. Here, we have an 8-bit integer that can only hold values in the range [0, 255]. The first two, 111 and 222, fit, but the last two, 333 and 444, are too large. NumPy quietly kept only the lowest 8 bits of these values, which correspond to 77 and 188, respectively. The lesson is that NumPy expects you to know what you're doing in regard to data types. Usually, this isn't an issue, but it is something to keep in mind.

2D Arrays

If a list turns into a 1D vector, we might suspect that a list of lists would turn into a 2D array. We'd be right:

```
>>> d = np.array([[1,2,3],[4,5,6],[7,8,9]])
>>> d.shape
(3, 3)
```

```
>>> d.size
9
>>> d
array([[1, 2, 3],
       [4, 5, 6],
       [7, 8, 9]])
```

We see that a list of three sublists is mapped to a 3×3 array (a matrix). Subscripts on NumPy arrays count from zero, so referencing d[1,2] above returns 6.

Zeros and Ones

Two particularly useful NumPy functions are np.zeros and np.ones. Both define arrays given a shape. The first initializes the array elements to zero, while the second initializes them to one. This is the primary way to create NumPy arrays from scratch:

```
>>> a = np.zeros((3,4), dtype="uint32")
>>> a[0,3] = 42
>>> a[1,1] = 66
>>> a
array([[ 0,  0,  0, 42],
       [ 0, 66,  0,  0],
       [ 0,  0,  0,  0]], dtype=uint32)
>>> b = 11*np.ones((3,1))
>>> b
array([[11.],
       [11.],
       [11.]])
```

The first argument is a tuple giving the size of each dimension. If we pass in a scalar, the resulting array is a 1D vector. Let's look at the definition of b. Here, we multiply the 3×1 array by a scalar (11). This causes each element of the array, which was initialized to 1.0, to be multiplied by 11.

Advanced Indexing

We saw simple array indexing in the examples above, where we indexed with a single value. NumPy supports more sophisticated array indexing. One type we'll use often is a single index that returns a complete subarray. Here's an example:

```
>>> a = np.arange(12).reshape((3,4))
>>> a
array([[ 0,  1,  2,  3],
       [ 4,  5,  6,  7],
       [ 8,  9, 10, 11]])
>>> a[1]
```

```
array([4, 5, 6, 7])
>>> a[1] = [44,55,66,77]
>>> a
array([[ 0,  1,  2,  3],
       [44, 55, 66, 77],
       [ 8,  9, 10, 11]])
```

This example introduces np.arange, which is the NumPy equivalent of
Python's range function. Notice the use of the reshape method to change the
12-element vector into a 3×4 matrix. Also, notice that a[1] returns the entire subarray, starting with the first index of the first dimension. This syntax
is short for a[1,:] where : means all elements of the given dimension. This
shorthand also works for assignments, as the next line shows.

The same syntax for indexing slices from a Python list works with Num-
Py. Here's what that looks like if we continue with the example above:

```
>>> a[:2]
array([[ 0,  1,  2,  3],
       [44, 55, 66, 77]])
>>> a[:2,:]
array([[ 0,  1,  2,  3],
       [44, 55, 66, 77]])
>>> a[:2,:3]
array([[ 0,  1,  2],
       [44, 55, 66]])
>>> b = np.arange(12)
>>> b
array([ 0,  1,  2,  3,  4,  5,  6,  7,  8,  9, 10, 11])
>>> b[::2]
array([ 0,  2,  4,  6,  8, 10])
>>> b[::3]
array([0, 3, 6, 9])
>>> b[::-1]
array([11, 10,  9,  8,  7,  6,  5,  4,  3,  2,  1,  0])
```

We see that a[:2] returns the first two rows with an implied : for the second dimension, as the following line shows. With our third command, we
get a subarray in two dimensions by taking the first two rows and first three
columns with a[:2,:3]. The examples with b show how to extract every other
or every third element. The last example is particularly handy: it uses a negative increment to reverse the dimension. The increment is −1 to reverse all
values. If it was −2, we'd get every other element of b in reverse order.

NumPy uses : to indicate all the elements along a specific dimension. It also allows ... (ellipsis) as a shorthand for "as many :s as needed." For example, let's define a three-dimensional (3D) array:

```
>>> a = np.arange(24).reshape((4,3,2))
>>> a
array([[[ 0,  1],
        [ 2,  3],
        [ 4,  5]],
       [[ 6,  7],
        [ 8,  9],
        [10, 11]],
       [[12, 13],
        [14, 15],
        [16, 17]],
       [[18, 19],
        [20, 21],
        [22, 23]]])
```

You can think of array a as a collection of four 3×2 matrices. To update the second of these matrices, you could use the following:

```
>>> a[1,:,:] = [[11,22],[33,44],[55,66]]
>>> a
array([[[ 0,  1],
        [ 2,  3],
        [ 4,  5]],
       [[11, 22],
        [33, 44],
        [55, 66]],
       [[12, 13],
        [14, 15],
        [16, 17]],
       [[18, 19],
        [20, 21],
        [22, 23]]])
```

Here, we specified the dimensions explicitly with : and showed that NumPy isn't picky: it knows that a list of lists matched the expected shape of the subarray and updated array a accordingly. We get the same effect by using the ellipsis as seen next.

```
>>> a[2,...] = [[99,99],[99,99],[99,99]]
>>> a
array([[[  0,   1],
        [  2,   3],
        [  4,   5]],
       [[ 11,  22],
        [ 33,  44],
        [ 55,  66]],
       [[ 99,  99],
        [ 99,  99],
        [ 99,  99]],
       [[ 18,  19],
        [ 20,  21],
        [ 22,  23]]])
```

We've now updated the third 3×2 subarray.

Reading and Writing to Disk

NumPy arrays can be written to and loaded from disk by using np.save and np.load, like so:

```
>>> a = np.random.randint(0,5,(3,4))
>>> a
array([[4, 2, 1, 3],
       [4, 0, 2, 4],
       [0, 4, 3, 1]])
>>> np.save("random.npy",a)
>>> b = np.load("random.npy")
>>> b
array([[4, 2, 1, 3],
       [4, 0, 2, 4],
       [0, 4, 3, 1]])
```

Here, we're using np.random.randint to create a random 3×4 integer array with values in the range 0 through 5. NumPy has extensive libraries for random numbers. We write array a to disk as random.npy. The .npy extension is necessary and will be added if we don't supply it. We then load the array back from disk using np.load.

We'll encounter other NumPy functions throughout the book. I'll explain them when they're first introduced. Let's move on now to a quick look at the SciPy library.

SciPy

SciPy adds a plethora of functionality to Python. It uses NumPy under the hood, so the two are often installed together. A full tutorial is available here: *https://docs.scipy.org/doc/scipy/reference/tutorial/index.html*.

In this book, we'll focus on the functions in the `scipy.stats` module. Start Python and try the following:

```
>>> import scipy
>>> scipy.__version__
'1.2.1'
```

This loads the SciPy module and verifies that the version number is at least what it should be. Any later version of SciPy should work just fine.

As a quick test, let's try the following:

```
>>> from scipy.stats import ttest_ind
>>> a = np.random.normal(0,1,1000)
>>> b = np.random.normal(0,0.5,1000)
>>> c = np.random.normal(0.1,1,1000)
>>> ttest_ind(a,b)
Ttest_indResult(statistic=-0.027161815649563964, pvalue=0.9783333836992686)
>>> ttest_ind(a,c)
Ttest_indResult(statistic=-2.295584443456226, pvalue=0.021802794508002675)
```

First, we load NumPy and then the `ttest_ind` function from SciPy's stats module. This function takes two sets of data, say test scores from two classes, and asks the question: do these sets of data have the same average value? Or, more accurately, it asks: how strongly can we believe that the same process generated these two sets of data? The *t-test* is a classic method for answering this question. One way to evaluate its result is to look at the *p-value*. You can think of a *p*-value as the probability that the two sets would have the measured difference in average value if they came from the same generating process. A probability near 1 means we have a lot of confidence that the two sets are from the same process.

The variables a, b, and c are 1D arrays where the values from the array (here 1,000) are extracted from Gaussian curves, also called *normal curves*. We'll get to these later, but for now, know that the numbers are pulled from a bell curve where values near the middle are more likely to be selected than those near the edges. The first two arguments to `normal` are the average value and the standard deviation, a measure of how spread out the bell curve is: the larger the standard deviation, the flatter, and wider, the curve.

For this example, we'd expect a and b to be very similar, as they both have an average value of 0.0, though slightly different bell curve shapes. However, c has an average value of 0.1. We hope the t-test picks up on this and tells us that we might want to be careful in believing a and c were generated by the same process.

The output of the ttest_ind function lists the *p*-value (pvalue). And, as we expected, comparing a and b returns a *p*-value of 0.98, meaning that the probability we'd see the difference between the averages of these two sets of data, given they came from the same generating process, is about 98 percent. However, when we compare a and c, we get a *p*-value of 2.7 percent (0.027). This means there is about a 3 percent chance we'd see the difference between a and c if they were generated by the same process. Therefore, we conclude that a and c are from different processes. We state, then, that the difference between these two datasets is *statistically significant*.

Historically, *p*-values less than 0.05 have been considered statistically significant. However, this threshold is arbitrary, and recent experience in replicating experiments, especially in the soft sciences, has led to a call for a stricter threshold. Using a *p*-value of 0.05 means you'll be wrong about 1 time in 20 (1/20 = 0.05), which is too generous a threshold. That said, a *p*-value close to 0.05 suggests that *something* is going on, and more investigation (and a larger dataset) is warranted.

Matplotlib

We'll use Matplotlib to generate graphs. Let's verify its 2D and 3D plotting abilities here. First, a simple 2D example:

```
>>> import numpy as np
>>> import matplotlib.pylab as plt
>>> x = np.random.random(100)
>>> plt.plot(x)
>>> plt.show()
```

This example loads NumPy, with which Matplotlib works best, and generates a vector, x, of 100 random values, $[0, 1)$, the output of np.random.random. We then use plt.plot to plot the vector and plt.show to display it. Matplotlib output is interactive. Play around with the plot to get familiar with how to use the plot window. For example, Figure 1-1 shows what the plot window looks like on Linux. Since the plot is random, you'll see a different sequence of values, but the controls on the window will be the same.

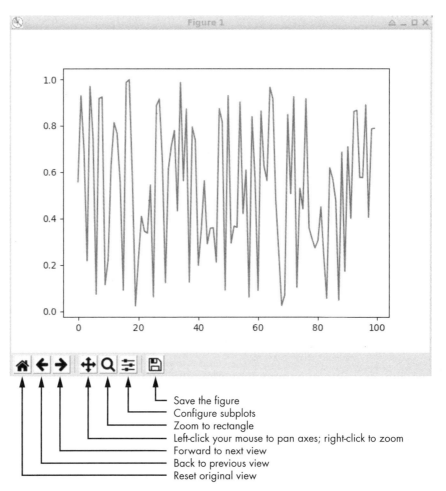

Figure 1-1: A sample Matplotlib plotting window

For 3D, give this a try:

```
>>> from mpl_toolkits.mplot3d import Axes3D
>>> import matplotlib.pylab as plt
>>> import numpy as np
>>> x = np.random.random(20)
>>> y = np.random.random(20)
>>> z = np.random.random(20)
>>> fig = plt.figure()
>>> ax = fig.add_subplot(111, projection='3d')
>>> ax.scatter(x,y,z)
>>> plt.show()
```

We first load the 3D axes toolkit, Matplotlib, and NumPy. Then, using NumPy, we generate three random vectors, $[0, 1)$. These are our 3D points. Using plt.figure and fig.add_subplot, we set up a 3D projection. The 111 is shorthand telling Matplotlib that we want a grid of 1×1 and that the current plot should go in index 1 of that grid. So, 111 means a single plot. The projection keyword gets the plot ready for 3D. Finally, the scatter plot is made, ax.scatter, and shown, plt.show. As with the 2D plot, the 3D plot is interactive. Grab and hold with the mouse to rotate the plot.

Scikit-Learn

The goal of this book is to cover the math of deep learning, not the implementation of deep learning. However, from time to time, it'll be helpful to look at a simple neural network model or two. In those cases, we'll make use of sklearn, in particular, the MLPClassifier class. Also, sklearn contains some useful tools for evaluating the performance of a model and for visualization of high-dimensional data.

As a quick example, let's build a simple neural network to classify small 8×8-pixel grayscale images of handwritten digits. This dataset is built into sklearn. Here's the code for the example:

```
import numpy as np
from sklearn.datasets import load_digits
from sklearn.neural_network import MLPClassifier

❶ d = load_digits()
digits = d["data"]
labels = d["target"]

N = 200
❷ idx = np.argsort(np.random.random(len(labels)))
x_test, y_test = digits[idx[:N]], labels[idx[:N]]
x_train, y_train = digits[idx[N:]], labels[idx[N:]]

❸ clf = MLPClassifier(hidden_layer_sizes=(128,))
clf.fit(x_train, y_train)

score = clf.score(x_test, y_test)
pred = clf.predict(x_test)
err = np.where(y_test != pred)[0]
print("score    : ", score)
print("errors:")
print(" actual   : ", y_test[err])
print(" predicted: ", pred[err])
```

We first import NumPy. From sklearn itself, we import the load_digits function to return the small digit image dataset and the MLPClassifier class to train a traditional neural network, that is, a multilayer perceptron. We then get the digit data and pull out the images and their associated labels, 0 . . . 9 ❶. The digit images are stored as 8 × 8 = 64-element vectors representing the image unraveled so the rows are laid end to end. The digits dataset includes 1,797 images, so digits is a 2D NumPy array with 1,797 rows, with 64 columns per row, and labels is a vector of 1,797 digit labels.

We randomize the order of the images, being careful to keep the right label with the right digit ❷ and extract train and test data (x_train, x_test) and labels (y_train, y_test). We'll set the first 200 digit images aside to use as test data and train the model with the remaining 1,597 images. This leaves us with approximately 160 images of each digit to train with and about 20 of each digit for testing.

Next, we build the model by creating an instance of MLPClassifier ❸. We'll take all the defaults and specify only the size of the single hidden layer, which has 128 nodes. The input vectors are 64 elements, so we double that for the hidden layer. There's no need to specify the output layer size explicitly; sklearn deduces it from the labels in y_train. Training the model is a simple call to clf.fit passing the training image vectors (x_train) and labels (y_train).

Training for a small dataset like this will take only a few seconds. When it's done, the learned weights and biases are in the model (clf). We first get the score, the overall accuracy (score), and then the actual model class label predictions on the test set (pred). Any errors are captured in err by looking for places where the actual label (y_test) does not match the prediction. We end by showing the actual class label and predicted label for the errors.

Each time we run this code, we'll get a different ordering of the digit data, which leads to a different train and test set. Additionally, neural networks are randomly initialized prior to training. So, we'll get a different result each time. The first time I ran this code, I had an overall score of 0.97 (97 percent) accuracy. Guessing would give an accuracy of about 10 percent, so we can say that the model has learned rather well.

Summary

In this chapter, we learned how to configure our working environment. I then introduced our suite of Python toolkits at a high level and supplied pointers on where to learn more. With the work environment secure and flourishing, the next chapter dives headfirst into probability theory.

2

PROBABILITY

Probability affects every aspect of our lives, but in reality, we're all pretty bad at it, as some of the examples in this chapter demonstrate. We need to study probability to get it right. And we need to get it right because deep learning deals extensively with ideas from probability theory. Probability appears everywhere, from the outputs of neural networks to how often different classes appear in the wild to the distributions used to initialize deep networks.

This chapter aims to expose you to the sorts of probability-related ideas and terms you'll frequently encounter in deep learning. We'll start with basic ideas about probability and introduce the notion of a random variable. We'll then jump to the rules of probability. These sections cover the basics that will put us in a position to talk about *joint and marginal probabilities*. You'll encounter those terms over and over again as you explore deep learning. Once you understand how to use joint and marginal probabilities, I'll explain the first of the two chain rules discussed in this book. The second is in Chapter 6 on differential calculus. We'll continue our study of probability in Chapter 3.

Basic Concepts

A *probability* is a number between zero and one that measures how likely something is to happen. If there's no chance the something will happen, its probability is zero. If it's absolutely certain that it will happen, its probability is one. We'll usually express probabilities this way, though in everyday use, people seem to dislike saying things like, "The chance of rain tomorrow is 0.25." Instead, we say, "The chance of rain tomorrow is 25 percent." In everyday speech, we convert the fractional probability to a percentage. We'll do the same in this chapter.

The previous paragraph used multiple words associated with probability: *likely*, *chance*, and *certainty*. This is fine in casual use, and even somewhat in deep learning, but when we need to be explicit, we'll stick with *probability* and express it numerically in the range zero to one, $[0, 1]$. The square brackets mean the upper and lower limit are included. If the limit isn't included in the range, a normal parenthesis is used. For example, the NumPy function np.random.random() returns a pseudorandom floating-point number in the range $[0, 1)$. So, it might return exactly zero, but it will never return exactly one.

Next, I'll introduce the foundational concepts of sample space, events, and random variables. I'll close with some examples of how humans are bad at probability.

Sample Space and Events

Put succinctly, a *sample space* is a discrete set or continuous range that represents all the possible outcomes of an event. An *event* is something that happens. Usually, it's the outcome of some physical process, like the flipping of a coin or the roll of a die. All the possible events we've grouped together are the sample space we're working with. Each event is a *sample* from the sample space, and the sample space represents *all* the possible events. Let's look at a few examples.

The possible outcomes of a coin flip are heads (H) or tails (T); thus, the sample space for a coin flip is the set $\{H, T\}$. The sample space for the roll of a standard die is the set $\{1, 2, 3, 4, 5, 6\}$ because, discounting the die perching itself on its edge, one of the six faces of the cube will be on top when the die stops moving. These are examples of discrete sample spaces. In deep learning, most sample spaces are continuous and they consist of floating-point numbers, not integers or elements of a set. For example, if a feature input to a neural network can take on any value in the range $[0, 1]$, then $[0, 1]$ is the sample space for that feature.

We can ask about the likelihood of certain events happening. For a coin, we can ask, what's the likelihood the coin will land heads up when flipped? Intuitively, assuming the coin isn't weighted so that one side is more likely to show up than the other, we say the likelihood of heads is 50 percent. The probability of getting heads is then 0.5 (50 percent as a percentage). We see that the probability of getting tails is also 0.5. Finally, since heads and tails are the only possible outcomes, we see that the sum of the probabilities over all possible results is $0.5 + 0.5 = 1.0$. Probabilities *always* sum to 1.0 over all possible values of the sample space.

What's the probability of rolling a four with a six-sided die? Again, there is no reason to favor one face over another, and only one of the six faces has four dots, so the probability is one out of six, $1/6 \approx 0.166666\ldots$ or about 17 percent.

Random Variables

Let's denote the outcome of a coin flip by a variable, X. X is what's called a *random variable*, a variable that takes on values from its sample space with a certain probability. Because here the sample space is discrete, X is a *discrete random variable*, which we denote with an uppercase letter. For the coin, the probability of X being heads equals the probability of X being tails, both 0.5. To write this formally, we use

$$P(X = \text{heads}) = P(X = \text{tails}) = 0.5$$

where P is universally used to indicate the probability of the event in parentheses for the specified random variable. A *continuous random variable* is a random variable from a continuous sample space, denoted with a lowercase letter, like x. We usually talk about the probability of the random variable being in some range of the sample space, not a particular real number. For example, if we use the NumPy `random` function to return a value in $[0, 1)$, we can ask: What's the probability that it will return a value in the range $[0, 0.25)$? Since any number is as likely to be returned as any other, we say that the probability of being in that range is 0.25 or 25 percent.

Humans Are Bad at Probability

We'll dive into the math of probability in the next section. But before then, let's look at two examples involving probability that show how bad humans can be at it. Both of these examples have stumped experts, not because the experts are somehow lacking, but because our intuitions about probability are often entirely incorrect, and even experts are thoroughly human.

The Monty Hall Dilemma

This problem is a particular favorite of mine, as it confuses even mathematicians with advanced degrees. The dilemma is taken from an old American game show called *Let's Make a Deal*. The original host of the show, Monty Hall, would select a member of the audience and show that person three large closed doors labeled 1, 2, and 3. Behind one of the doors was a new car. Behind the remaining two doors were joke prizes, such as a live goat.

The contestant was asked to pick a door. Then, Hall would ask that one of the doors the contestant *didn't* pick be opened, naturally one that didn't have a car behind it. After the audience stopped laughing at whatever joke prize was behind that door, Hall would ask the contestant if they wanted to keep the originally selected door, or if they would rather change their selection to the remaining door. The dilemma is simply that: do they keep their original guess, or do they switch to the remaining door?

If you want to think about it for a while, please do. Put the book down, walk around, get out a pencil and some paper, make notes, then, when you have a solution (or give up), read on....

Here's the right answer: change doors. If you do, you'll win the car 2/3 of the time. If you don't, you'll only win the car 1/3 of the time, as that's the probability of selecting the correct door initially: one correct choice out of three.

When Marilyn vos Savant presented this problem in her *Parade* magazine column in 1990 and stated that the correct solution is to change doors, she was flooded with letters, many from mathematicians, some angry, insisting she was wrong. She wasn't. One way to see that she was right is to use a computer program to simulate the game. We won't develop the code for one here, but it isn't too hard. If you write one and run it, you'll see the probability of winning when changing doors converges on 2/3 as the number of simulated games increases. However, we can also use common sense and basic ideas about probability to see the solution.

First, if we don't change doors, we know we have a 1/3 probability of winning the car. Now, consider what can happen when we change doors. If we change doors, the only way we can *lose* is if we happened to select the correct door in the first place. Why? Suppose we initially chose one of the joke prize doors instead. Hall, who knows full well which door the car is behind, will never open the door with the car. Since we selected one of the joke doors already, he's forced to choose the remaining joke door and open it for us, thereby ensuring the car is behind the only remaining door. If we switch doors, we win. Since there are two doors without the car, our chance of selecting the wrong door initially is 2/3. However, we just saw that if we choose the wrong door initially and switch when given the opportunity, we'll win the car. Therefore, we have a 2/3 chance of winning the car by changing our guess. The 1/3 probability of losing by changing our initial guess is, of course, the case where we initially selected the correct door.

Cancer or Not?

This example is found in several popular books about probability and statistics (for instance, *More Damned Lies and Statistics*, by Joel Best [UC Press, 2004], and *The Drunkard's Walk*, by Leonard Mlodinow [Pantheon, 2008]). It's based on an actual study. The task is to determine the probability that a woman in her 40s has breast cancer if she has a positive mammogram. Note that the numbers that follow might have been accurate when the study was conducted, but they may not be valid now. Please consider them only as an example.

We are told the following:

1. The probability that a randomly selected woman in her 40s has breast cancer is 0.8 percent (8 out of 1,000).
2. The probability that a woman with breast cancer will have a positive mammogram is 90 percent.
3. The probability that a woman *without* breast cancer will have a positive mammogram is 7 percent.

A woman comes to the clinic and is screened. The mammogram is positive. What's the probability, based on what we've been told, that she actually has breast cancer?

From #1 above, we know that if we select 1,000 women in their 40s at random, 8 of them will have breast cancer (on average). Therefore, of those 8, 90 percent of them (#2 above) will have a positive mammogram. This means 7 women with cancer will have a positive mammogram because $8 \times 0.9 = 7.2$. This leaves 992 of the original 1,000 who don't have breast cancer. From #3 above, $992 \times 0.07 = 69.4$, so 69 women without breast cancer will also have a positive mammogram, giving a total of $7 + 69 = 76$ positive mammograms, of which 7 are actual cancer and 69 are false-positive results. Therefore, the probability that a positive mammogram indicates cancer is 7 out of 76 or $7/76 = 0.092$—about 9 percent.

The median estimate that doctors presented with this problem gave was a probability of cancer of around 70 percent, with over one-third giving an estimate of 90 percent. Probabilities are hard for humans, even for those with a lot of training. The doctors' mistake wasn't properly accounting for the probability of a randomly selected woman in her 40s having breast cancer. We'll see in Chapter 3 how to calculate this result using Bayes' theorem, which does take this probability into account.

For now, let's switch from intuition to mathematical formality.

The Rules of Probability

Let's get started with the basic rules of probability. These are foundational rules that we'll need for the remainder of the chapter and beyond. We'll learn about the probability of events, the sum rule for probabilities, and what we mean by a conditional probability. After that, the product rule will let us tackle the birthday paradox. In the birthday paradox, we'll see how to calculate the minimum number of people to have together in a room such that the probability of at least two of them sharing a birthday exceeds 50 percent. The answer is fewer than you might think.

Probability of an Event

We mentioned earlier that the sum of all the probabilities for a sample space is one. This means that the chance of any event from the sample space is always less than or equal to one, since the event came from the sample space, and the sample space encompasses all possible events. This implies, for any event A,

$$0 \leq P(A) \leq 1 \tag{2.1}$$

and, for all events A_i in the sample space,

$$\sum_i P(A_i) = 1 \tag{2.2}$$

where Σ (sigma) means to sum over the expression on the right for each of the i's. Think of a for loop in Python with the expression on the right as the body of the loop.

If we roll a six-sided die, we intuitively (and correctly) understand that the probability of getting any value is the same: one out of six possibilities, or $1/6$. Therefore, Equation 2.1 tells us that $P(1)$, the probability of rolling a one, is between zero and one. This is true since $0 \leq \frac{1}{6} \leq 1$. Furthermore, Equation 2.2 tells us that the sum of the probabilities of all events in the sample space must be one. This is also true for the six-sided die, since $P(1) = P(2) = P(3) = P(4) = P(5) = P(6) = \frac{1}{6}$ and $\frac{1}{6} + \frac{1}{6} + \frac{1}{6} + \frac{1}{6} + \frac{1}{6} + \frac{1}{6} = 1$.

If the probability of an event happening is $P(A)$, then the probability that event A *does not* happen is

$$P(\bar{A}) = 1 - P(A) \tag{2.3}$$

with $P(\bar{A})$ read as "not A." $P(\bar{A})$ is known as the *complement* of A. You'll sometimes see $P(\bar{A})$ written as $P(\neg A)$ using $\neg$, the logical symbol for "not."

Equation 2.3 comes from Equation 2.1 and Equation 2.2 because the probability of an event is less than one and the probability of any event from the sample space happening is one, so the probability of events that aren't A happening must be one minus the probability of event A happening.

For example, when rolling a die, the probability of getting a value in $[1, 6]$ is one, but the probability of getting a four is $1/6$. So, the chance of *not* rolling a four is all the probability that remains when the chance of rolling a four is removed,

$$P(\bar{4}) = 1 - P(4) = 1 - \frac{1}{6} = \frac{5}{6} = 0.8333\ldots$$

meaning we have an 83 percent chance of not rolling a four.

What if we roll two dice and sum them? The sample space is the set of integers from 2 through 12. However, each sum is not equally likely in this case, a situation that's at the core of the casino game craps, for example. We calculate the probabilities of each sum by enumerating all the ways they can happen. By counting the ways events can happen and dividing by the total number of events, we can determine the probability. Table 2-1 shows all the possible ways to generate each sum.

Table 2-1: The Number of Combinations of Two Dice Leading to Different Sums

Sum	Combinations	Count	Probability
2	1 + 1	1	0.0278
3	1 + 2, 2 + 1	2	0.0556
4	1 + 3, 2 + 2, 3 + 1	3	0.0833
5	1 + 4, 2 + 3, 3 + 2, 4 + 1	4	0.1111
6	1 + 5, 2 + 4, 3 + 3, 4 + 2, 5 + 1	5	0.1389
7	1 + 6, 2 + 5, 3 + 4, 4 + 3, 5 + 2, 6 + 1	6	0.1667
8	2 + 6, 3 + 5, 4 + 4, 5 + 3, 6 + 2	5	0.1389
9	3 + 6, 4 + 5, 5 + 4, 6 + 3	4	0.1111
10	4 + 6, 5 + 5, 6 + 4	3	0.0833
11	5 + 6, 6 + 5	2	0.0556
12	6 + 6	1	0.0278
		36	1.0000

In Table 2-1, there are 36 possible combinations of the two dice. We see that the most likely sum is 7, since six combinations add to 7. The least likely are 2 and 12; there's only one way to get either. If there are six ways to get a sum of 7, then the probability of a 7 is "6 out of 36," or $6/36 \approx 0.1667$. We'll return to Table 2-1 later in the next chapter when we discuss probability distributions and Bayes' theorem. Table 2-1 illustrates a general rule: if we can enumerate the sample space, then we can calculate the probabilities of specific events.

As a final example, if you flip three coins simultaneously, what is the probability of getting no heads, one head, two heads, or three heads? We can enumerate the possible outcomes and see. We get the following:

Heads	Combinations	Count	Probability
0	TTT	1	0.125
1	HTT, THT, TTH	3	0.375
2	HHT, HTH, THH	3	0.375
3	HHH	1	0.125
		8	1.000

From this table, we claim that the probability of getting one or two heads in three coin flips is the same: 37.5 percent. Let's test this with a bit of code:

```python
import numpy as np
N = 1000000
M = 3
heads = np.zeros(M+1)
for i in range(N):
    flips = np.random.randint(0,2,M)
    h, _ = np.bincount(flips, minlength=2)
    heads[h] += 1
```

```
prob = heads / N
print("Probabilities: %s" % np.array2string(prob))
```

The code runs 1,000,000 tests (N) simulating the flip of three coins (M). The number of times each test ends up with 0, 1, 2, or 3 heads is stored in heads. Each test selects three values in $[0, 1]$ (flips) and counts how many heads (a zero) show up. We use np.bincount for this and throw away the number of tails. The number of heads is then tallied, and the next set of flips happens.

When all N simulations are complete, we convert the number of heads to probabilities by dividing by the number of simulations run (prob). Finally, we print the corresponding probabilities. For zero, one, two, or three heads, a single run returned the following:

```
Probabilities: [0.125236, 0.3751, 0.37505, 0.124614]
```

These are quite close to the probabilities we calculated above, so we have confidence that we're correct.

Sum Rule

We'll start with a definition: two events A and B are said to be *mutually exclusive* if they can't both happen; either one or the other happens. For example, a coin flip is either heads or tails; it can't be heads *and* tails. Mutually exclusive events mean if event A happens, event B is excluded, and vice versa. Additionally, if the probabilities of two events happening are completely unrelated, meaning the probability of A is unaffected by whether B has happened, we say the two events are *independent*.

The sum rule is concerned with the probability of more than one mutually exclusive event happening. It tells us the probability of either event happening. For example, what's the probability of rolling a four or a five with a standard die? We know the probability of rolling a four is $1/6$, as is the probability of rolling a five. Since the events are mutually exclusive, we can intuit that the probability of getting a four *or* a five is their sum, since four and five as outcomes are both parts of the sample space, and either one or the other happens or neither happens. So, we get the following:

$$P(A \text{ or } B) = P(A \cup B) = P(A) + P(B) \text{ (for mutually exclusive events)} \quad (2.4)$$

Here $\cup$ means "or" or "union." You'll see $\cup$ often. For a standard die, the probability of rolling a four or a five is $\frac{1}{6} + \frac{1}{6} = \frac{1}{3}$, or about 33 percent.

The sample space of two coin flips is {HH, HT, TH, TT}; therefore, this is the probability of getting two heads or two tails:

$$P(\text{HH or TT}) = P(\text{HH}) + P(\text{TT}) = \frac{1}{4} + \frac{1}{4} = \frac{1}{2}$$

There's more to the sum rule, but before we can see it, we need to consider the product rule.

Product Rule

The sum rule tells us about the probability of events A or B happening. The product rule tells us the probability of events A *and* B:

$$P(A \text{ and } B) = P(A \cap B) = P(A)P(B) \tag{2.5}$$

Here $\cap$ means "and" or "intersection."

If events A and B are mutually exclusive, we will immediately see that $P(A \cap B) = 0$ because if event A happens with probability $P(A)$, then event B's probability is $P(B) = 0$, and their product is also zero. The same is true if event B happens; then $P(A) = 0$.

Not all events are mutually exclusive, of course. For example, assume 80 percent of the people in the world have brown eyes and 50 percent are female. What's the probability of a randomly selected person being a female with brown eyes? Let's use the product rule,

$$P(\text{female, brown-eyed}) = P(\text{female})P(\text{brown-eyed}) = 0.5(0.8) = 0.4$$

to see there is a 40 percent chance of a randomly selected person being a brown-eyed female.

The product rule makes sense if we think about it a bit. Calculating the fraction of people, which is the probability, who are female won't change the fraction of those females who are brown-eyed. One event, being female, has no impact on the other event, being brown-eyed.

The product rule isn't limited to only two events. Consider the following. According to insurance companies, the probability of being struck by lightning in any given year, if you live in the US, is about $1/1{,}222{,}000$, or 0.000082 percent. What's the probability of being a brown-eyed female and being struck by lightning in any given year, assuming you live in the US? Again, we can use the product rule:

$$P(\text{female, brown-eyed, lightning}) = P(\text{female})P(\text{brown-eyed})P(\text{lightning})$$

$$= 0.5(0.8)(0.00000082)$$

$$= 0.00000033 = 0.000033\%$$

The population of the United States is about $331{,}000{,}000$, of which 0.000033 percent are brown-eyed females who'll be struck by lightning this year: 109 people, by our calculation above. According to the US National Weather Service, about 270 people will be struck by lightning in a given year. As we saw above, 40 percent of those people will be brown-eyed females, which yields $270(0.4) = 108$. So, our calculation is entirely believable.

Sum Rule Revisited

We stated above that there's more to the sum rule. Let's see now what we were missing above. Equation 2.4 gives us the sum rule for mutually exclusive events A and B. What if the events aren't mutually exclusive? In that case, the sum rule needs to be modified:

$$P(A \text{ or } B) = P(A) + P(B) - P(A \text{ and } B) \qquad (2.6)$$

Let's look at an example.

An archaeologist has discovered a small cache of 20 ancient coins. He notes that 12 of the coins are Roman and 8 are Greek. He also notes that 6 of the Roman coins and 3 of the Greek coins are silver. The remaining coins are bronze. What's the probability of selecting a silver or Roman coin from the cache?

If we believe that silver and Roman are mutually exclusive, we'd be tempted to say the following:

$$P(\text{silver or Roman}) = P(\text{silver}) + P(\text{Roman})$$

$$= \frac{9}{20} + \frac{12}{20} \quad (\textit{This is wrong!})$$

However, the sum of the two probabilities is $\frac{21}{20} = 1.05$, and we *can't* have a probability greater than one. Something is amiss.

The problem is that there are Roman coins in the cache that are made of silver. We counted them twice—once in $P(\text{silver})$ and again in $P(\text{Roman})$—so now we need to subtract them from the overall sum. There are six silver Roman coins. So, the probability of being a silver Roman coin is $P(\text{silver and Roman}) = \frac{6}{20}$. Subtracting that part, we see that the probability of picking a silver coin or a Roman coin is 75 percent:

$$P(\text{silver or Roman}) = P(\text{silver}) + P(\text{Roman}) - P(\text{silver and Roman})$$

$$= \frac{9}{20} + \frac{12}{20} - \frac{6}{20}$$

$$= \frac{15}{20} = 0.75$$

As with the sum rule, there's more to the product rule, and we'll get to that shortly. But first, let's use the product rule to see if we can solve the birthday paradox.

The Birthday Paradox

On average, how many people do we need together in a room to have a higher than 50 percent chance that two of them share the same birthday? This problem is known as the *birthday paradox*. Let's see if we can use our knowledge of the product rule for probability to see what the solution is.

We'll ignore leap years and state that there are 365 days in a year. Intuitively, we see that the probability of randomly selected people sharing a birthday is one day (the shared birthday) out of 365 possible birthdays in a year. The sample space is 365 days, and the shared birthday is the one day in common. So, we get the following:

$$P(\text{share a birthday}) = \frac{1}{365} \approx 0.00274$$

They either share a birthday or they don't: $1 - 1/365 = 365/365 - 1/365$ $= 364/365$. So we get the following:

$$P(\text{do not share a birthday}) = 1 - \frac{1}{365} = \frac{364}{365} \approx 0.9973$$

Of the 365 days in a year, there is one possible match, leaving 364 days that don't match.

A 0.3 percent chance of randomly selected people sharing a birthday is pretty low. It means if you randomly choose pairs of people and ask if they share a birthday, on average you'll get three matches in a thousand—not terribly likely.

For our calculation, we'll look at things the other way. We're looking for the number of people we need together so that the probability of *no* two people sharing a birthday is below 50 percent.

We know the probability that two randomly selected people *don't* share a birthday: $\frac{364}{365}$. Therefore, if we select two pairs of people at random, the probability that both pairs do not share a birthday is the following:

$$P(\text{neither share a birthday}) = P(\text{do not share})P(\text{do not share})$$

$$= \left(\frac{364}{365}\right)\left(\frac{364}{365}\right)$$

$$= 0.9945 = 99.45\%$$

Here, we're using the product rule. Similarly, with three people, (A, B, C), we can form three different pairs, (A, B), (A, C), and (B, C), so we calculate the following:

$$P(\text{no shared birthday}) = \left(\frac{364}{365}\right)\left(\frac{364}{365}\right)\left(\frac{364}{365}\right)$$

For n comparisons, here's the probability that none share a birthday:

$$P(\text{no shared birthday}) = \left(\frac{364}{365}\right)^n \tag{2.7}$$

Our task is to find the minimum number of comparisons, n, leading to a probability of no shared birthday < 50 percent, where n is a function of the number of people in the room, m. Why less than 50 percent? Because if we find an n leading to a less than 50 percent probability that there is no shared birthday, the probability that there *is* a shared birthday must then be > 50 percent.

If you pick three people at random, there are three pairs of people to check to see if they share a birthday. If you have four people, there are six pairs. So, the larger the group of people, the more pairs there are. Can we find a rule that maps the number of people, m, to the number of pairs to compare, n? If we have that, we can find the smallest m leading to an n where the probability of Equation 2.7 is < 50 percent.

When we have a set of m unique objects, like people in a room, and we select pairs of them, how many different pairs can we select? In other words, how many combinations of m things are there when taken two at a time? The formula to calculate the number of combinations of m things taken k at a time is this:

$$C(m, k) = \binom{m}{k} = \frac{m!}{k!(m-k)!}$$

You'll sometimes hear this referred to as "m choose k," where, for us, $k = 2$. Let's find the number of comparisons we need, n, and use the number of combinations of things taken two at a time to find an m leading to at least n comparisons.

A straightforward loop in Python locates the n we need:

```
for n in range(300):
    if ((364/365)**n < 0.5):
        print(n)
        break
```

We're told $n = 253$. So, we need to make, on average, 253 comparisons, 253 pairs of people, to have a greater than 50 percent chance that one of those pairs shares a birthday. The final step is to find how many combinations of m people taken two at a time are at least 253. A bit of brute force trial and error tells us this:

$$\binom{23}{2} = \frac{23!}{2!(23-2)!}$$

$$= \frac{23!}{2!(21!)}$$

$$= \frac{23(22)}{2}$$

$$= 253$$

We need $m = 23$ people on average to have a higher than 50 percent chance at least two of them share a birthday. All thanks to the product rule.

Is our result reliable or just sleight of hand? Some code can tell us. First, let's verify via simulation that the probability of randomly picking two people who share a birthday is 0.3 percent:

```
match = 0
for i in range(100000):
    a = np.random.randint(0,364)
    b = np.random.randint(0,364)
    if (a == b):
        match += 1
print("Probability of a random match = %0.6f" % (match/100000,))
```

The code simulates 100,000 random pairs of people, where the random integer in $[0, 364]$ represents the person's birthday. If the two random birthdays match, match is incremented. After all simulations run, we print the probability. A run of this code produced the following, making our assertion of a 0.3 percent chance believable:

```
Probability of a random match = 0.003100
```

What about the number of people to get a > 50 percent chance of sharing a birthday? Here, we have two loops. The first is over the number of people in the room (m), and the second is over the number of simulations for that many people in the room (n). In code, that looks like this:

```
for m in range(2,31):
    matches = 0
    for n in range(100000):
        match = 0
        b = np.random.randint(0,364,m)
        for i in range(m):
            for j in range(m):
                if (i != j) and (b[i] == b[j]):
                    match += 1
        if (match != 0):
            matches += 1
    print("%2d %0.6f" % (m, matches/100000))
```

We let m range over 2 to 30 people. For each set of m people, we run 100,000 simulations. For each simulation, we pick a set of birthdays for each person in the room (b) and then compare each person with every other person to see if there's a matching birthday. If there is, we increment match. If we had at least one match, we increment matches and move to the next simulation. Finally, when all of the simulations for the current number of people in the room are complete, we print the probability of at least a single match.

If we run the code and plot the output, we get Figure 2-1, where the dashed line is 50 percent. The first point above the dashed line is 23 people, precisely as we calculated.

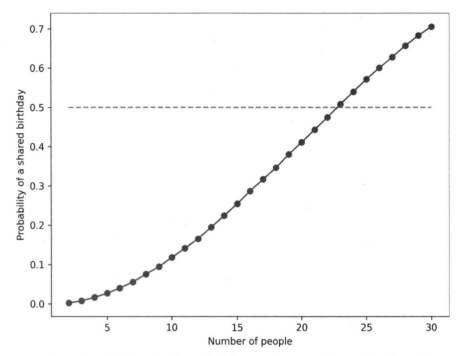

Figure 2-1: The probability of a shared birthday as a function of the number of people in a room

It's always satisfying to see the simulation line up with the math.

Conditional Probability

Consider a bag of 10 marbles: 8 red and 2 blue. We know if we pick a marble at random from the bag, we have a 2 out of 10, or 20 percent, chance of picking a blue marble. Say we pick a blue marble. After admiring its pretty shade of blue, we put it back in the bag, shake the bag, and pull out another marble. What's our chance of picking a blue marble a second time? Again, there are 2 blue marbles and 10 total, so it's still 20 percent.

If the fact that event *A* happened (here, picking a blue marble that we then returned to the bag) hasn't affected the probability of a future event *B*, the two are independent events. Our chance of picking a blue marble a second time is in no way affected by the fact that we previously chose a blue marble. The same is true for a coin flip. The fact that we've landed heads up four times in a row has nothing to do with the probability of getting tails on the next flip, assuming it's a fair coin, i.e., it's not weighted on one side or two-headed (or two-tailed).

Now, consider an alternate scenario. We still have a bag with eight red and two blue marbles. We pick a marble—say it's red this time—and because we like the color, we keep the marble and put it aside. Now, we pick another marble from the bag. What's the probability of picking another red marble? Here, things have changed. There are nine marbles now, and seven of them are red. So, our chance of picking a second red marble is now 7 of 9, or

78 percent. The possibility of choosing a red marble initially was 8 of 10, or 80 percent. The fact that event A happened, picking a red marble we then kept, has altered the probability of a second event. The two events are no longer independent. The probability of the second event was changed by the first event happening. Notationally, we write $P(B|A)$ to mean the probability of event B *given* event A has happened. This is a *conditional probability* because it's conditional on event A happening.

Here's where we update the product rule. The version in Equation 2.5 assumes that the two events are independent, like being female and having brown eyes. If we have a dependent situation, the rule becomes

$$P(A \text{ and } B) = P(B|A)P(A) \tag{2.8}$$

meaning the probability of the two events both happening is the product of the probability of one given the other has happened and the probability of the other.

Looking back at our marble example above, we calculated the probability of picking a red marble after already picking and keeping a red marble to be 7 of 9, or about 78 percent. That's $P(B|A)$. For $P(A)$, we need the probability of picking a red marble initially, which we said was 80 percent. Therefore, the probability of picking a red marble that we keep, A, and picking a red marble on a second draw, B, is 62 percent:

$$P(A \text{ and } B) = P(B|A)P(A)$$

$$= \left(\frac{7}{9}\right)\left(\frac{8}{10}\right)$$

$$= 0.6222$$

If two events are mutually exclusive, $P(B|A) = P(A|B) = 0$. If events A and B are independent, then $P(A|B) = P(A)$ and $P(B|A) = P(B)$ because the conditional event happening or not has no influence on the later event.

Finally, note that typically $P(B|A) \neq P(A|B)$, and confusing the two conditional probabilities is a common and often serious error. As we'll see in Chapter 3, something called Bayes' theorem gives the proper relationship between the conditional probabilities. We'll encounter conditional probability again when discussing the chain rule for probability.

Total Probability

If our sample space is separated into disjoint regions, B_i (B_1, B_2, etc.) so that the totality of the sample space is covered by the collection of B_is and the B_is don't overlap, we can calculate the probability of an event over all the partitions as the following:

$$P(A) = \sum_i P(A|B_i)P(B_i)$$

Here $P(A|B_i)$ is the probability of A given partition B_i and $P(B_i)$ is the probability of partition B_i, which is the amount of the sample space that B_i represents. In this view, $P(A)$ is the *total probability* of A over the partitions, B_i. Let's see an example of how to use this law.

You have three cities, Kish, Kesh, and Kuara, and their populations are 2,000, 1,000, and 3,000, respectively. Additionally, the percentages of people with blue eyes in these cities are 12 percent, 3 percent, and 21 percent, respectively. We want to know the probability that a randomly selected person from among the cities has blue eyes. The cities' populations affect things, as the probability of having blue eyes varies by city and the cities vary in population. To find $P(\text{blue})$, we use the total probability:

$$P(\text{blue}) = P(\text{blue}|\text{Kish})P(\text{Kish})$$

$$+ P(\text{blue}|\text{Kesh})P(\text{Kesh})$$

$$+ P(\text{blue}|\text{Kuara})P(\text{Kuara})$$

Here $P(\text{blue}|\text{Kish})$ is the probability of having blue eyes given you live in Kish, and $P(\text{Kish})$ is the probability of living in Kish, and so on.

We know the necessary quantities to find the total probability. The probability of blue eyes per city is given above, and the probability of living in each city is found from its population and the total population of the three cities:

$$P(\text{Kish}) = 2000/6000 = \frac{1}{3}$$

$$P(\text{Kesh}) = 1000/6000 = \frac{1}{6}$$

$$P(\text{Kuara}) = 3000/6000 = \frac{1}{2}$$

Therefore, $P(\text{blue})$ is

$$P(\text{blue}) = 0.12\left(\frac{1}{3}\right) + 0.03\left(\frac{1}{6}\right) + 0.21\left(\frac{1}{2}\right) = 0.15$$

meaning there is a 15 percent chance a randomly selected inhabitant of the three cities will have blue eyes. Note the sum of the probabilities for selecting the cities: $P(\text{Kish}) + P(\text{Kesh}) + P(\text{Kuara}) = 1$. This must be the case for the partitioning of the total sample space, all the inhabitants of the cities, to be covered by the partitioning into cities.

Joint and Marginal Probability

The *joint probability* of two variables, $P(X = x, Y = y)$, is the probability that random variable X will have the value x at the same time random variable

Y is y. We've already seen an example of a joint probability. When we use "and" when calculating a probability, we're calculating a joint probability. A joint probability is the probability of multiple conditions being true at the same time, which is "and." The *marginal probability* is what we get when we calculate the probability of one or more of those conditions without caring about the value of the others; in other words, the probability of a subset of the random variables in the "and."

In this section, we'll examine joint and marginal probabilities using simple tables. We'll then introduce the chain rule for probability. This rule lets us break down a joint probability into the product of smaller joint probabilities and conditional probabilities.

Joint Probability Tables

According to Colour Blind Awareness (*http://www.colourblindawareness.org/*), approximately 1 in 12 men and 1 in 200 women are color-blind. The difference comes from the fact that the affected gene is on the X chromosome, requiring a woman to inherit the recessive gene from both her mother and father. A man need only inherit the gene from one parent.

Pretend we survey 1,000 people. We can count the number of people who are male and color-blind, female and color-blind, male and not color-blind, and female and not color-blind. We do this and arrange the data in a table like so:

	Color-blind	Not color-blind	
Male	42	456	498
Female	3	499	502
	45	955	1000

Tables like these are known as *contingency tables*. The tallied data is in the center 2 × 2 numerical portion of the table. The rightmost column is the sum across the rows, and the final row is the sum of the columns. The sum of the final row or column is in the last cell and, by necessity, sums to the 1,000 people we surveyed.

We can turn the contingency table into a table of probabilities by dividing each cell by 1,000, the number of people surveyed. Doing this gives us the following:

	Color-blind	Not color-blind	
Male	0.042	0.456	0.498
Female	0.003	0.499	0.502
	0.045	0.955	1.000

The table is now a joint probability table. With it, we can look up the probability of being male and color-blind. Notationally, we write

$$P(\text{sex} = \text{male}, \text{color-blind} = \text{yes}) = 0.042$$

and, similarly, we see that

$$P(\text{sex} = \text{female}, \text{color-blind} = \text{no}) = 0.499$$

Using the joint probability table, we can predict what we would expect to measure given a random sample of people. For example, if we have a sample of 20,000 people, then, based on our table, we'll expect to find about $20000(0.042) = 840$ color-blind men and about $20000(0.003) = 60$ color-blind women.

What if we wanted to know the probability of being color-blind regardless of sex? For that, we sum the probabilities along the color-blind column and see that there is a 4.5 percent chance that a randomly selected person is color-blind. Likewise, summing along the row gives us an estimated probability of being female as 50.2 percent. We do need to bear in mind that our table was built from a sample of only 1,000 people. You might guess that if we had instead sampled 100,000 people, our split between male and female would be closer to 50/50, and you'd be right.

Calculating the probability of being color-blind or female from the joint probability table is calculating a marginal probability. In the first case, we summed along the column to remove the effect of sex, whereas in the second case, we summed along the row to remove the effect of color-blindness.

Mathematically, we get the marginal probabilities by summing across the variables we don't want. If we have a joint probability table for two variables, like the example above, we get the marginal probabilities by summing:

$$P(X = x) = \sum_i P(X = x, Y = y_i)$$

$$P(Y = y) = \sum_i P(X = x_i, Y = y)$$

Using the table above, we can write

$$P(Y = \text{color-blind}) = P(X = \text{male}, Y = \text{color-blind})$$

$$+ P(X = \text{female}, Y = \text{color-blind})$$

where we sum across sex to remove its effect. Now, let's explore another table, one with three variables.

Sometime during the night of April 14, 1912, the RMS *Titanic* sank in the North Atlantic on its maiden voyage from England to New York City. Based on a sample of 887 people who were on board the *Titanic*, we can generate Table 2-2 showing the joint probability for three variables: survival, sex, and cabin class.

Table 2-2: Joint Probability Table for *Titanic* Passengers

		Cabin1	Cabin2	Cabin3
Dead	Male	0.087	0.103	0.334
	Female	0.003	0.007	0.081
Alive	Male	0.051	0.019	0.053
	Female	0.103	0.079	0.081

Let's use Table 2-2 to calculate some probabilities. Note, we'll use the values in Table 2-2, which are accurate to three decimals. As a result, the overall numbers will be slightly off from the probabilities we'd calculate from the counts, but doing this makes the link between the table and the equations more concrete.

First, we can read directly from the table for specific triplets of survived, sex, and cabin class. Here's an example:

$$P(\text{dead, male, cabin3}) = 0.334$$

This means the probability of a randomly selected passenger being a man who was in a third-class cabin and didn't survive is 33 percent. What about men in first class? That's in the table too:

$$P(\text{dead, male, cabin1}) = 0.087$$

This means that a selected passenger has a 9 percent chance of being a man in first class who died. We can see that class differences, in cabins and society, mattered quite a bit.

Let's use the table to calculate some other joint and marginal probabilities. First, what's the probability of not surviving? To find it, we need to sum over sex and cabin:

$$P(\text{dead}) = P(\text{dead}, M, 1) + P(\text{dead}, M, 2) + P(\text{dead}, M, 3) \quad (2.9)$$

$$+ P(\text{dead}, F, 1) + P(\text{dead}, F, 2) + P(\text{dead}, F, 3)$$

$$= 0.087 + 0.103 + 0.334 + 0.003 + 0.007 + 0.081$$

$$= 0.615$$

Here we've introduced a shorthand notation for male/female (M/F) and cabin class $(1, 2, 3)$.

Let's calculate the probability of not surviving *given* the passenger was male, $P(\text{dead}|M)$. To do this, we look back to Equation 2.8, remembering that the "and" implies a joint probability. We rewrite Equation 2.8 to solve for $P(B|A)$:

$$P(B|A) = \frac{P(A, B)}{P(A)}$$

This is sometimes used to define the conditional probability in the first place. Note, $P(A, B)$ means $P(A$ and $B)$—both are joint probabilities. Using this form, the probability of not surviving given the passenger is male is

$$P(\text{dead}|M) = \frac{P(\text{dead}, M)}{P(M)}$$

where $P(\text{dead}, M)$ is the joint probability of being dead and male, and $P(M)$ is the probability of being male.

We need to be careful when thinking about the probabilities. $P(\text{dead}, M)$ is not the probability of not surviving if the passenger is male. Instead, it's the probability of a randomly selected passenger being a male who didn't survive. What we want is $P(\text{dead}|M)$, which is the probability of not surviving given a passenger was male.

To get $P(\text{dead}, M)$, we need to sum over cabin class:

$$P(\text{dead}, M) = P(\text{dead}, M, 1) + P(\text{dead}, M, 2) + P(\text{dead}, M, 3) \qquad (2.10)$$

$$= 0.087 + 0.103 + 0.334$$

$$= 0.524$$

To get $P(M)$, we sum over survival and cabin class:

$$P(M) = P(\text{dead}, M, 1) + P(\text{dead}, M, 2) + P(\text{dead}, M, 3) \qquad (2.11)$$

$$+ P(\text{alive}, M, 1) + P(\text{alive}, M, 2) + P(\text{alive}, M, 3)$$

$$= 0.087 + 0.103 + 0.334 + 0.051 + 0.019 + 0.053$$

$$= 0.647$$

To finally calculate $P(\text{dead}|M)$:

$$P(\text{dead}|M) = \frac{P(\text{dead}, M)}{P(M)} = \frac{0.524}{0.647} = 0.810$$

This tells us that 81 percent of the male passengers didn't survive.

A similar calculation, shown below, tells us the probability of being female and surviving:

$$P(\text{alive}|F) = \frac{P(\text{alive}, F)}{P(F)} = \frac{0.263}{0.354} = 0.743$$

We see that women were far more likely to survive than men. Here's one instance where the phrase "women and children first" was actually the case. I leave it as an exercise for you to calculate the individual probabilities in $P(\text{alive}|F)$.

We've calculated $P(\text{dead}, M)$, the probability of being a male who did not survive; $P(M)$, the probability of being male; and $P(\text{dead}|M)$, the probability of not surviving given being a male. Let's do one more calculation based on Table 2-2. Let's find $P(\text{dead or } M)$, the probability of not surviving, or being male.

Equation 2.6 tells us that this is the probability:

$$P(\text{dead or } M) = P(\text{dead}) + P(M) - P(\text{dead}, M)$$

$$= 0.615 + 0.647 - 0.524$$

$$= 0.738$$

If we look at Equation 2.9 and Equation 2.11, we see that both have the same terms, the very terms summed in Equation 2.10. This is why we must subtract $P(\text{dead}, M)$ from the calculation of $P(\text{dead or } M)$ to avoid double-counting.

To summarize, then:

- The joint probability is the probability of two or more random variables having a specific set of values. The joint probability is often represented as a table.

- The marginal probability for a random variable is found by summing over all the possible values of the other random variables.

The product rule with a conditional probability tells us how to calculate the joint probability given a conditional probability and an unconditional probability when we have two random variables. Let's now see how to use the chain rule for probability to generalize that idea.

Chain Rule for Probability

Equation 2.8 tells us how to calculate the joint probability for two random variables in terms of the conditional probability. By using the *chain rule for probability*, we can expand Equation 2.8 and calculate the joint probability for more than two random variables.

In its generic form, the chain rule for the joint probability of n random variables is as follows:

$$P(X_n, X_{n-1}, \ldots, X_1) = \prod_{i=1}^{n} P\left(X_i \middle| \bigcap_{j=1}^{i-1} X_j \right) \tag{2.12}$$

Here $\bigcap$ is used to indicate "and" for joint probabilities. Equation 2.12 looks impressive, but it's not hard to follow, as we'll see with a few examples. I need to use $\bigcap$ in the equation for the joint part of the conditional probabilities, but in the examples, I'll use a comma, and you'll see the pattern quickly enough.

Here's how the chain rule breaks up a joint probability with three random variables:

$$P(X, Y, Z) = P(X|Y, Z)P(Y, Z)$$

$$= P(X|Y, Z)P(Y|Z)P(Z)$$

The first line says the probability of X, Y, and Z is the product of the probability of X given Y and Z and the probability of Y and Z. This is Equation 2.8 with X for B and Y, Z for A. The second line applies the chain rule to $P(Y, Z)$ to get $P(Y|Z)P(Z)$. The rule can be applied in sequence, like a chain, hence the name.

What about a joint probability with four random variables? We get the following:

$$P(A, B, C, D) = P(A|B, C, D)P(B, C, D)$$

$$= P(A|B, C, D)P(B|C, D)P(C, D)$$

$$= P(A|B, C, D)P(B|C, D)P(C|D)P(D)$$

Let's work through an example that uses the chain rule. Say we're very social and have 50 people at our party. Four of the 50 people have been to Boston in the fall. We pick three people at random. What's the probability that *none* of them have been to Boston in the fall?

We'll use A_i to indicate the event of a person who has not been to Boston in the fall. Therefore, what we want to find is $P(A_3, A_2, A_1)$, the probability of three people all of whom have not been to Boston in the fall. The chain rule lets us break this probability up like so:

$$P(A_3, A_2, A_1) = P(A_3|A_2, A_1)P(A_2|A_1)P(A_1)$$

We can work with the right-hand side of this equation intuitively. Look at $P(A_1)$. This is the probability of picking a random person in the room who has not been to Boston in the fall. Four of the people have, so 46 have not, and we see that $P(A_1) = 46/50$. Once we have a person picked, we need to know the probability of selecting a second person from the remaining 49, that's $P(A_2|A_1) = 45/49$. There are only 49 people left, and we haven't selected one of the four who has been to Boston in the fall. Finally, we have two people selected, so there are 48 people in the room, 44 of whom have not been to Boston in the fall. This means $P(A_3|A_2, A_1) = 44/48$.

We are now ready to answer our initial question. The probability of selecting three people from the room with none of them having been to Boston in the fall is as follows:

$$P(A_3, A_2, A_1) = P(A_3|A_2, A_1)P(A_2|A_1)P(A_1)$$

$$= \left(\frac{44}{48}\right)\left(\frac{45}{49}\right)\left(\frac{46}{50}\right)$$

$$= 0.7745$$

That's slightly more than 77 percent.

We can check if our calculation is reasonable by simulating many draws of three people. This is the code we need:

```
nb = 0
N = 100000
for i in range(N):
    s = np.random.randint(0,50,3)
    fail = False
    for t in range(3):
        if (s[t] < 4):
            fail = True
    if (not fail):
        nb += 1
print("No Boston in the fall = %0.4f" % (nb/N,))
```

We'll run 100,000 simulations. Every time we select three people out of 50 who haven't been to Boston in the fall, we'll increment nb. We simulate selecting three people by choosing three random integers in the range $[0, 50)$ and putting them in s. Then, we look at each of the three integers, asking if any are less than four. If any are, we say we selected a person who has been to Boston and set fail to True. If none of the three integers are less than four, we were successful with this simulation. When done, we print the fraction of simulations that did sample three people who never went to Boston in the fall.

Running this code produced

```
No Boston in the fall = 0.7780
```

which is close enough to our calculated value to give us confidence that we found the correct answer.

Summary

This chapter introduced the fundamentals of probability. We explored basic concepts of probability, including sample spaces and random variables. We followed with some examples of how poor humans can be at probability. After that, we considered the rules of probability, with examples. The rules led us to joint and marginal probabilities and finally to the chain rule for probabilities.

The next chapter continues our tour of probability, starting with probability distributions and how to sample from them and ending with Bayes' theorem, which shows us correct way to compare conditional probabilities.

3

MORE PROBABILITY

Chapter 2 introduced us to basic concepts of probability. In this chapter, we'll continue our exploration of probability by focusing on two essential topics often encountered in deep learning and machine learning: probability distributions and how to sample from them, and Bayes' theorem. Bayes' theorem is one of the most important concepts in probability theory, and it has produced a paradigm shift in the way many researchers think about probability and how to apply it.

Probability Distributions

A probability distribution can be thought of as a function that generates values on demand. The values generated are random—we don't know which one will appear—but the likelihood of any value appearing follows a general form. For example, if we roll a standard die many times and tally how many times each number comes up, we expect that in the long run, each number is equally likely. Indeed, that's the entire point of making the die in the first place. Therefore, the probability distribution of the die is known as a *uniform distribution*, since each number is equally likely to appear. We can imagine other distributions favoring one value or range of values over others, like a weighted die that might come up as six suspiciously often.

Deep learning's primary reason for sampling from a probability distribution is to initialize the network before training. Modern networks select the initial weights and sometimes biases from different distributions, most notably uniform and normal. The uniform distribution is familiar to us, and I'll discuss the normal distribution, a continuous distribution, later.

I'll present several different kinds of probability distributions in this section. Our focus is to understand the shape of the distribution and to learn how to draw samples from it using NumPy. I'll start with histograms to show you that we can often treat histograms as approximations of a probability distribution. Then I'll discuss common *discrete probability distributions*. These are distributions returning integer values, like 3 or 7. Lastly, I'll switch to continuous distributions yielding floating-point numbers, like 3.8 or 7.592.

Histograms and Probabilities

Take a look at Table 3-1, which we saw in Chapter 2.

Table 3-1: The Number of Combinations of Two Dice Leading to Different Sums (Copied from Table 2-1)

Sum	Combinations	Count	Probability
2	1 + 1	1	0.0278
3	1 + 2, 2 + 1	2	0.0556
4	1 + 3, 2 + 2, 3 + 1	3	0.0833
5	1 + 4, 2 + 3, 3 + 2, 4 + 1	4	0.1111
6	1 + 5, 2 + 4, 3 + 3, 4 + 2, 5 + 1	5	0.1389
7	1 + 6, 2 + 5, 3 + 4, 4 + 3, 5 + 2, 6 + 1	6	0.1667
8	2 + 6, 3 + 5, 4 + 4, 5 + 3, 6 + 2	5	0.1389
9	3 + 6, 4 + 5, 5 + 4, 6 + 3	4	0.1111
10	4 + 6, 5 + 5, 6 + 4	3	0.0833
11	5 + 6, 6 + 5	2	0.0556
12	6 + 6	1	0.0278
		36	1.0000

It shows how two dice add to different sums. Don't look at the actual values; look at the shape the possible combinations make. If we chop off the last two columns, turn the table to the left, and replace each sum with an "X," we should see something like the following.

```
                              ×
                    ×    ×    ×
               ×    ×    ×    ×    ×
          ×    ×    ×    ×    ×    ×    ×
     ×    ×    ×    ×    ×    ×    ×    ×    ×
×    ×    ×    ×    ×    ×    ×    ×    ×    ×    ×
─────────────────────────────────────────────────
2    3    4    5    6    7    8    9   10   11   12
```

You can see that there's a definite shape and symmetry to the number of ways to arrive at each sum. This kind of plot is called a *histogram*. A histogram is a plot tallying the number of things that fall into discrete bins. For Table 3-1, the bins are the numbers 2 through 12. The tally is a possible way to get that sum. Histograms are often represented as bar graphs, usually vertical bars, though they need not be. Table 3-1 is basically a horizontal histogram. How many bins are used in the histogram is up to the maker. If you use too few, the histogram will be blocky and may not reveal necessary detail because interesting features have all been lumped into the same bin. Use too many bins, and the histogram will be sparse, with many bins having no tallies.

Let's generate some histograms. First, we'll randomly sample integers in [0,9] and count how many of each integer we get. The code for this is straightforward:

```
>>> import numpy as np
>>> n = np.random.randint(0,10,10000)
>>> h = np.bincount(n)
>>> h
array([ 975, 987, 987, 1017, 981, 1043, 1031, 988, 1007, 984])
```

We first set n to an array of 10,000 integers in [0, 9]. We then use np .bincount to count how many of each digit we have. We see that this run gave us 975 zeros and 984 nines. If the NumPy pseudorandom generator is doing its job, we expect, on average, to have 1,000 of each digit in a sample of 10,000 digits. We expect some variation, but most values are close enough to 1,000 to be convincing.

The counts above tell us how many times each digit appeared. If we divide each bin of a histogram by the total of all the bins, we change from simple counts to the probability of that bin appearing. For the random digits above, we get the probabilities with

```
>>> h = h / h.sum()
>>> h
array([0.0975, 0.0987, 0.0987, 0.1017, 0.0981, 0.1043, 0.1031, 0.0988,
       0.1007, 0.0984])
```

which tells us that each digit did appear with a probability of about 0.1, or 1 out of 10. This trick of dividing histogram values by the sum of the counts in the histogram allows us to estimate probability distributions from samples. It also tells us the likelihood of particular values appearing when sampling from whatever process generated the data used to make the histogram. You should note that I said we could *estimate* the probability distribution from a set of samples drawn from it. The larger the number of samples, the closer the estimated probability distribution will be to the actual population distribution generating the samples. We will never get to the actual population distribution, but given the limit of an infinite number of samples, we can get as close as we need to.

Histograms are frequently used to look at the distribution of pixel values in an image. Let's make a plot of the histogram of the pixels in two images. You can find the code in the file ricky.py. (I won't show it here, as it doesn't add to the discussion.) The images used are two example grayscale images included with SciPy in scipy.misc. The first shows people walking up stairs (ascent), and the second is the face of a young raccoon (face), as shown in Figure 3-1.

Figure 3-1: People ascending (left) and "Ricky" raccoon (right)

Figure 3-2 provides a plot of the histograms for each image, as probabilities. It shows two very different distributions of gray level values in the images. For the raccoon face, the distribution is more spread out and flatter, while the ascent image has a spike right around gray level 128 and a few bright pixels. The distributions tell us that if we pick a random pixel in the face image, we're most likely to get one around gray level 100, but an arbitrary pixel in the ascent image will, with high relative likelihood, be closer to gray level 128.

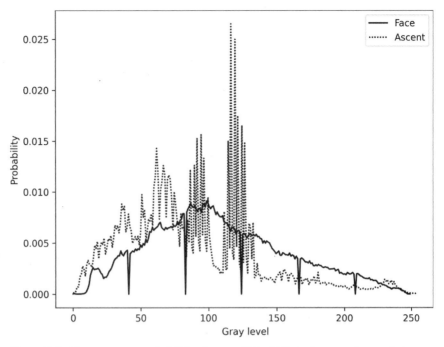

Figure 3-2: Histograms as probabilities for two 512×512-pixel grayscale sample images

Again, histograms tally the counts of how many items fall into the predefined bins. We saw for the images that the histogram as probability distribution tells us how likely we are to get a particular gray level value if we select a random pixel. Likewise, the probability distribution for the random digits in the example before that tells us the probability of getting each digit when we ask for a random integer in the range [0,9].

Histograms are discrete representations of a probability distribution. Let's take a look at the more common discrete distributions now.

Discrete Probability Distributions

We've already encountered the most common discrete distribution several times: it's the uniform distribution. That's the one we get naturally by rolling dice or flipping coins. In the uniform distribution, all possible outcomes are equally likely. A histogram of a simulation of a process drawing from a uniform distribution is flat; all outcomes show up with more or less the same frequency. We'll see the uniform distribution again when we look at continuous distributions. For now, think dice.

Let's look at a few other discrete distributions.

The Binomial Distribution

Perhaps the second most common discrete distribution is the *binomial distribution*. This distribution represents the expected number of events happening in a given number of trials if each event has a specified probability. Mathematically, the probability of k events happening in n trials if the probability of the event happening is p can be written as

$$P(X = k) = \binom{n}{k} p^k (1 - p)^{n-k}$$

For example, what's the probability of getting three heads in a row when flipping a fair coin three times? From the product rule, we know the probability is

$$P(HHH) = \left(\frac{1}{2}\right)\left(\frac{1}{2}\right)\left(\frac{1}{2}\right) = \frac{1}{8} = 0.125$$

Using the binomial formula, we get the same answer by calculating

$$P(HHH) = \binom{3}{3}(0.5)^3(1 - 0.5)^{3-3} = 0.125$$

So far, not particularly helpful. However, what if the probability of the event isn't 0.5? What if we have an event, say the likelihood of a person winning *Let's Make a Deal* by not changing doors, and we want to know the probability that 7 people out of 13 will win by not changing their guess? We know the probability of winning the game without changing doors is 1/3—that's p. We then have 13 trials (n) and 7 winners (k). The binomial formula tells us the likelihood is

$$P(X = 7) = \binom{13}{7}\left(\frac{1}{3}\right)^7 (1 - \frac{1}{3})^{13-7} = 0.0689$$

and, if the players *do* switch doors,

$$P(X = 7) = \binom{13}{7}\left(\frac{2}{3}\right)^7 (1 - \frac{2}{3})^{13-7} = 0.1378$$

The binomial formula gives us the probability of a given number of events in a given number of trials for a specified probability per event. If we fix n and p and vary k, $0 \leq k \leq n$, we get the probability for each k value.

This gives us the distribution. For example, let $n = 5$ and $p = 0.3$, then $0 \le k \le 5$ with the probability for each k value as

$$P(X = 0) = \binom{5}{0}(0.3)^0(1 - 0.3)^{5-0} = 0.1681$$

$$P(X = 1) = \binom{5}{1}(0.3)^1(1 - 0.3)^{5-1} = 0.3601$$

$$P(X = 2) = \binom{5}{2}(0.3)^2(1 - 0.3)^{5-2} = 0.3087$$

$$P(X = 3) = \binom{5}{3}(0.3)^3(1 - 0.3)^{5-3} = 0.1323$$

$$P(X = 4) = \binom{5}{4}(0.3)^4(1 - 0.3)^{5-4} = 0.0283$$

$$P(X = 5) = \binom{5}{5}(0.3)^5(1 - 0.3)^{5-5} = 0.0024$$

Allowing for rounding, this sums to 1.0, as we know it must because the sum of probabilities over an entire sample space is always 1.0. Notice that we calculate all the possible values for the binomial distribution when $n = 5$. Collectively, this specifies the *probability mass function (pmf)*. The probability mass function tells us the probability associated with all possible outcomes.

The binomial distribution is parameterized by n and p. For $n = 5$ and $p = 0.3$, we see from the results above that a random sample from such a binomial distribution will return 1 most often—some 36 percent of the time. How can we draw samples from a binomial distribution? In NumPy, we need only call the binomial function in the random module:

```
>>> t = np.random.binomial(5, 0.3, size=1000)
>>> s = np.bincount(t)
>>> s
array([159, 368, 299, 155,  17,   2])
>>> s / s.sum()
array([0.159, 0.368, 0.299, 0.155, 0.017, 0.002])
```

We pass binomial the number of trials (5) and the probability of success for each trial (0.3). We then ask for 1,000 samples from a binomial distribution with these parameters. Using np.bincount, we see that the most commonly returned value was indeed 1, as we calculated above. By using our histogram summation trick, we get a probability of 0.368 for selecting a 1—close to the 0.3601 we calculated.

The Bernoulli Distribution

The *Bernoulli distribution* is a special case of the binomial distribution. In this case, we fix $n = 1$, meaning there's only one trial. The only values we can sample are 0 or 1; either the event happens, or it doesn't. For example, with $p = 0.5$, we get

```
>>> t = np.random.binomial(1, 0.5, size=1000)
>>> np.bincount(t)
array([496, 504])
```

This is reasonable, since a probability of 0.5 means we're flipping a fair coin, and we see that the proportion of heads or tails is roughly equal.

If we change to $p = 0.3$, we get

```
>>> t = np.random.binomial(1, 0.3, size=1000)
>>> np.bincount(t)
array([665, 335])
>>> 335/1000
0.335
```

Again, close to 0.3, as we expect to see.

Use samples from a binomial distribution when you want to simulate events with a known probability. With the Bernoulli form, we can sample binary outcomes, 0 or 1, where the likelihood of the event need not be that of the flip of a fair coin, 0.5.

The Poisson Distribution

Sometimes, we don't know the probability of an event happening for any particular trial. Instead, we might know the average number of events that happen over some interval, say of time. If the average number of events that happen over some time is λ (lambda), then the probability of k events happening in that interval is

$$P(k) = \frac{\lambda^k e^{-\lambda}}{k!}$$

This is the *Poisson distribution*, and it's useful to model events like radioactive decay or the incidence of photons on an X-ray detector over some period of time. To sample events according to this distribution, we use poisson from the random module. For example, assume over some time interval there are five events on average ($\lambda = 5$). What sort of probability distribution do we get using the Poisson distribution? In code,

```
>>> t = np.random.poisson(5, size=1000)
>>> s = np.bincount(t)
>>> s
array([  6,  36,  83, 135, 179, 173, 156, 107,  58,  40,  20,   4,   2,
         0,   0,   1])
```

```
>>> t.max()
15
>>> s = s / s.sum()
>>> s
array([0.006, 0.036, 0.083, 0.135, 0.179, 0.173, 0.156, 0.107, 0.058,
       0.04 , 0.02 , 0.004, 0.002, 0.   , 0.   , 0.001])
```

Here, we see that, unlike the binomial distribution, which could not select more than n events, the Poisson distribution can select numbers of events that exceed the value of λ. In this case, the largest number of events in the time interval was 15, which is three times the average. You'll find that the most frequent number of events is right around the average of five, as you might expect, but significant deviations from the average are possible.

The Fast Loaded Dice Roller

What if we need to draw samples according to an arbitrary discrete distribution? Earlier, we saw some histograms based on images. In that case, we could sample from the distribution represented by the histogram by picking pixels in the image at random. But what if we wanted to sample integers according to arbitrary weights? To do this, we can use the new Fast Loaded Dice Roller of Saad, et al.[1]

The *Fast Loaded Dice Roller (FLDR)* lets us specify an arbitrary discrete distribution and then draw samples from it. The code is in Python and freely available. (See *https://github.com/probcomp/fast-loaded-dice-roller/*.) I'll show how to use the code to sample according to a generic distribution. I recommend downloading just the fldr.py and fldrf.py files from the GitHub repository instead of running setup.py. Additionally, edit the .fldr import lines in fldrf.py to remove the "." so they read

```
from fldr import fldr_preprocess_int
from fldr import fldr_s
```

Using FLDR requires two steps. The first is to tell it the particular distribution you want to sample from. You define the distribution as ratios. (For our purposes, we'll use actual probabilities, meaning our distribution will always add up to 1.0.) This is the preprocessing step, which we only need to do once for each distribution. After that, we can draw samples. An example will clarify:

```
>>> from fldrf import fldr_preprocess_float_c
>>> from fldr import fldr_sample
>>> x = fldr_preprocess_float_c([0.6,0.2,0.1,0.1])
>>> t = [fldr_sample(x) for i in range(1000)]
```

---

1. Feras A. Saad, Cameron E. Freer, Martin C. Rinard, and Vikash K. Mansinghka, "The Fast Loaded Dice Roller: A Near-Optimal Exact Sampler for Discrete Probability Distributions," in AISTATS 2020: Proceedings of the 23rd International Conference on Artificial Intelligence and Statistics, *Proceedings of Machine Learning Research* 108, Palermo, Sicily, Italy, 2020.

```
>>> np.bincount(t)
array([598, 190, 108, 104])
```

First, we import the two FLDR functions we need: `fldr_preprocess_float` `_c` and `fldr_sample`. Then we define the distribution using a list of four numbers. Four numbers imply samples will be integers in $[0, 3]$. However, unlike a uniform distribution, here we're specifying we want zero 60 percent of the time, one 20 percent of the time, and two and three 10 percent of the time each. The information that FLDR needs to sample from the distribution is returned in x.

Calling `fldr_sample` returns a single sample from the distribution. Notice two things: first, we need to pass x in, and second, FLDR doesn't use NumPy, so to draw 1,000 samples, we use a standard Python list comprehension. The 1,000 samples are in the list, t. Finally, we generate the histogram and see that nearly 60 percent of the samples are zero and slightly more than 10 percent are three, as we intended.

Let's use the histogram of the raccoon face image we used earlier to see if FLDR will follow a more complex distribution. We'll load the image, generate the histogram, convert it to a probability distribution, and use the probabilities to set up FLDR. After that, we'll draw 25,000 samples from the distribution, compute the histogram of the samples, and plot that histogram along with the original histogram to see if FLDR follows the actual distribution we give it. The code we need is

```
from scipy.misc import face
im = face(True)
b = np.bincount(im.ravel(), minlength=256)
b = b / b.sum()
x = fldr_preprocess_float_c(list(b))
t = [fldr_sample(x) for i in range(25000)]
q = np.bincount(t, minlength=256)
q = q / q.sum()
```

Running this code leaves us with b, a probability distribution from the histogram of the face image, and q, the distribution created from 25,000 samples from the FLDR distribution. Figure 3-3 shows us a plot of the two distributions.

The solid line in Figure 3-3 is the probability distribution we supplied to `fldr_preprocess_float_c` representing the distribution of gray levels (intensities) in the raccoon image. The dashed line is the histogram of the 25,000 samples from this distribution. As we can see, they follow the requested distribution with the sort of variation we expect from such a small number of samples. As an exercise, change the number of samples from 25,000 to 500,000 and plot the two curves. You'll see that they're now virtually on top of each other.

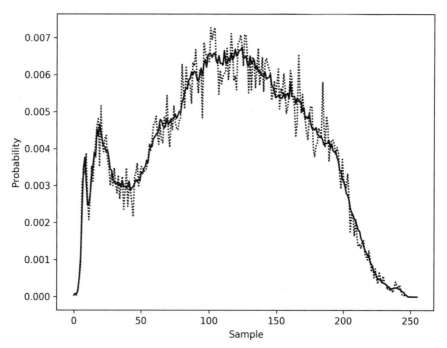

Figure 3-3: Comparing the Fast Loaded Dice Roller distribution (dashed) to the distribution generated from the SciPy face image (solid)

Discrete distributions generate integers with specific likelihoods. Let's leave them now and consider continuous probability distributions, which return floating-point values instead.

Continuous Probability Distributions

I haven't discussed continuous probabilities yet in this chapter. In part, not doing so was to make the concepts behind probability easier to follow. A continuous probability distribution, like a discrete one, has a particular shape. However, instead of assigning a probability to a specific integer value, as we saw above, the probability of selecting a particular value from a continuous distribution is zero. The probability of a specific value, a real number, is zero because there are an infinite number of possible values from a continuous distribution; this means no particular value can be selected. Instead, what we talk about is the probability of selecting values in a specific range of values.

For example, the most common continuous distribution is the uniform distribution over $[0, 1]$. This distribution returns *any* real number in that range. Although the probability of returning a specific real number is zero, we can talk about the probability of returning a value in a range, such as $[0, 0.25]$.

Consider again the uniform distribution over $[0, 1]$. We know that the sum of all the individual probabilities from zero to one is 1.0. So, what is the probability of sampling a value from this distribution and having that value be in the range $[0, 0.25]$? All values are equally likely, and all add to 1.0, so we must have a 25 percent chance of returning a value in $[0, 0.25]$. Similarly, we have a 25 percent chance of returning a value in $[0.75, 1]$, as that also covers $1/4$ of the possible range.

When we talk about summing infinitely small things over a range, we're talking about integration, the part of calculus that we won't cover in this book. Conceptually, however, we can understand what's happening if we think about a discrete distribution in the limit where the number of values it can return goes to infinity, and we're summing the probabilities over some range.

We also can think about this graphically. Figure 3-4 shows the continuous probability distributions I'll discuss.

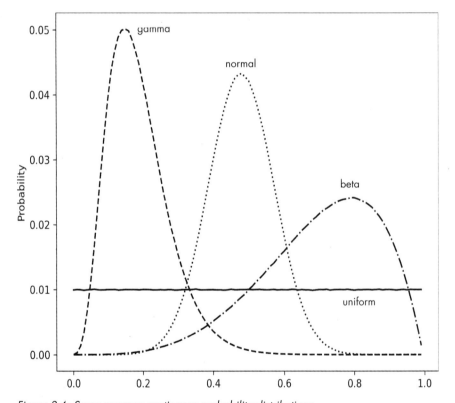

Figure 3-4: Some common continuous probability distributions

To get the probability of sampling a value in some range, we add up the area under the curve over that range. Indeed, this is precisely what integration does; the integration symbol ($\int$) is nothing more than a fancy "S" for *sum*. It's the continuous version of Σ for summing discrete values.

The distributions in Figure 3-4 are the most common ones you'll encounter, though there are many others useful enough to be given names.

All of these distributions have associated *probability density functions (pdfs)*, closed-form functions that generate the probabilities that sampling from the distribution will give. I generated the curves in Figure 3-4 instead using the code in the file continuous.py. The curves are estimates of the probability density functions, and I created them from the histogram of a large number of samples. I did so intentionally to demonstrate that the NumPy random functions sampling from these distributions do what they claim.

Pay little attention to the x-axis in Figure 3-4. The distributions have different ranges of output; they're scaled here to fit all of them on the graph. The important thing to notice is their shapes. The uniform distribution is, well, uniform over the entire range. The normal curve, also frequently called a *Gaussian* or a *bell curve*, is the second most common distribution used in deep learning. For example, the He initialization strategy for neural networks samples initial weights from a normal distribution.

The code generating the data for Figure 3-4 is worth considering, as it shows us how to use NumPy to get samples:

```
N = 10000000
B = 100
t = np.random.random(N)
u = np.histogram(t, bins=B)[0]
u = u / u.sum()
t = np.random.normal(0, 1, size=N)
n = np.histogram(t, bins=B)[0]
n = n / n.sum()
t = np.random.gamma(5.0, size=N)
g = np.histogram(t, bins=B)[0]
g = g / g.sum()
t = np.random.beta(5,2, size=N)
b = np.histogram(t, bins=B)[0]
b = b / b.sum()
```

NOTE *We're using the classic NumPy functions here, not the newer Generator-based functions. NumPy updated the pseudorandom number code in recent versions, but the overhead of using the new code will detract from what we want to see here. Unless you're very serious about pseudorandom number generation, the older functions, and the Mersenne Twister pseudorandom number generator they're based on, will be more than adequate.*

To make the plots, we first use 10 million samples from each distribution (N). Then, we use 100 bins in the histogram (B). Again, the x-axis range when plotting isn't of interest here, only the shapes of the curves.

The uniform samples use random, a function we've seen before. Passing the samples to histogram and applying the "divide by the sum" trick creates the probability curve data (u). We repeat this process for the Gaussian (normal), Gamma (gamma), and Beta (beta) distributions as well.

You'll notice that `normal`, `gamma`, and `beta` accept arguments. These distributions are parameterized; their shape is altered by changing these parameters. For the normal curve, the first parameter is the mean (μ), and the second is the standard deviation (σ). Some 68 percent of the normal curve lies within one standard deviation of the mean, $[\mu - \sigma, \mu + \sigma]$. The normal curve is ubiquitous in math and nature, and one could write an entire book on it alone. It's always symmetric around its mean value. The standard deviation controls how wide or narrow the curve is.

The gamma distribution is also parameterized. It accepts two parameters: the shape (k) and the scale (θ). Here, $k = 5$, and the scale is left at its default value of $\theta = 1$. As the shape increases, the gamma distribution becomes more and more like a Gaussian, with a bump that moves toward the center of the distribution. The scale parameter affects the horizontal size of the bump.

Likewise, the beta distribution uses two parameters, a and b. Here, $a = 5$ and $b = 2$. If $a > b$, the hump of the distribution is on the right; if reversed, it is on the left. If $a = b$, the beta distribution becomes the uniform distribution. The flexibility of the beta distribution makes it quite handy for simulating different processes, as long as you can find a and b values approximating the probability distribution you want. However, depending on the precision you require, the new Fast Loaded Dice Roller we saw in the previous section might be a better option in practice if you have a sufficiently detailed discrete distribution approximation of the continuous distribution.

Table 3-2 shows us the probability density functions for the normal, gamma, and beta distributions. An exercise for the reader is to use these functions to recreate Figure 3-4. Your results will be smoother still than the curves in the figure. You can calculate the $B(a, b)$ integral in Table 3-2 by using the function `scipy.special.beta`. For $\Gamma(k)$, see `scipy.special.gamma`. Additionally, if the argument to the Γ function is an integer, $\Gamma(n + 1) = n!$, so $\Gamma(5) = \Gamma(4 + 1) = 4! = 24$.

Table 3-2: The Probability Density Functions for the Normal, Gamma, and Beta Distributions

normal	$p(x) = \dfrac{1}{\sqrt{2\pi\sigma^2}} e^{-(x-\mu)^2/\sigma^2}$
gamma	$p(x) = x^{k-1} \dfrac{e^{-x/\theta}}{\theta^k \Gamma(k)}, \quad \Gamma(k) = \displaystyle\int_0^\infty t^{k-1} e^{-t} dt$
beta	$p(x) = \dfrac{1}{B(a,b)} x^{a-1}(1-x)^{b-1}, \quad B(a,b) = \displaystyle\int_0^1 t^{a-1}(1-t)^{b-1} dt$

If you're interested in ways to sample values from these distributions, my book *Random Numbers and Computers* (Springer, 2018) discusses these distributions and others in more depth than we can provide here, including implementations in C for generating samples from them. For now, let's examine one of the most important theorems in probability theory.

Central Limit Theorem

Imagine we draw N samples from some distribution and calculate the mean value, m. If we repeat this exercise many times, we'll get a set of mean values, $\{m_0, m_1, \ldots\}$, each from a set of samples from the distribution. It doesn't matter if N is the same each time, but N shouldn't be too small. The rule of thumb is that N should be at least 30 samples.

The *central limit theorem* states that the histogram or probability distribution generated from this set of sample means, the m's, will approach a Gaussian in shape regardless of the shape of the distribution the samples were drawn from in the first place.

For example, this code

```
M = 10000
m = np.zeros(M)
for i in range(M):
    t = np.random.beta(5,2,size=M)
    m[i] = t.mean()
```

creates 10,000 sets of samples from a beta distribution, Beta(5,2), each with 10,000 samples. The mean of each set of samples is stored in m. If we run this code and plot the histogram of m, we get Figure 3-5.

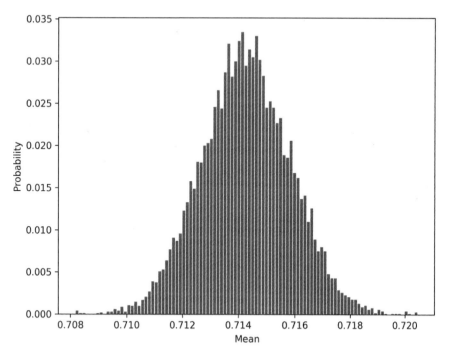

Figure 3-5: The distribution of mean values of 10,000 sets of samples of 10,000 from Beta(5,2)

The shape of Figure 3-5 is decidedly Gaussian. Again, the shape is a consequence of the central limit theorem and does not depend on the shape of the underlying distribution. Figure 3-5 tells us that the sample means from many sets of samples from Beta(5,2) themselves have a mean of about 0.714. The mean of the sample means (m.mean()) is 0.7142929 for one run of the code above.

There's a formula to calculate the mean value of a Beta distribution. The population mean value of a Beta(5,2) distribution is known to be $a/(a + b) = 5/(5 + 2) = 5/7 = 0.714285$. The mean of the plot in Figure 3-5 is a measurement of the true population mean, of which the many means from the Beta(5,2) samples are only estimates.

Let's explain this again to really follow what's going on. For any distribution, like the Beta(5,2) distribution, if we draw N samples, we can calculate the mean of those samples, a single number. If we repeat this process for many sets of N samples, each with its own mean, and we make a histogram of the distribution of the means we measured, we'll get a plot like Figure 3-5. That plot tells us that all of the many sample means are themselves clustered around a mean value. The mean value of the means is a measure of the population mean. It's the mean we'd get if we could draw an infinite number of samples from the distribution. If we change the code above to use the uniform distribution, we'll get a population mean of 0.5. Similarly, if we switch to a Gaussian distribution with a mean of 11, the resulting histogram will be centered at 11.

Let's prove this claim again but this time with a discrete distribution. Let's use the Fast Loaded Dice Roller to generate samples from a lopsided discrete distribution using this code:

```
from fldrf import fldr_preprocess_float_c
from fldr import fldr_sample
z = fldr_preprocess_float_c([0.1,0.6,0.1,0.1,0.1])
m = np.zeros(M)
for i in range(M):
    t = np.array([fldr_sample(z) for i in range(M)])
    m[i] = t.mean()
```

Figure 3-6 shows the discrete distribution (top) and the corresponding distribution of the sample means (bottom).

From the probability mass function, we can see that the most frequent value we expect from the sample is 1, with a probability of 60 percent. However, the tail on the right means we'll also get values 2 through 4 about 30 percent of the time. The weighted mean of these is $0.6(1) + 0.1(2) + 0.1(3) + 0.1(4) = 1.5$, which is precisely the mean of the sample distribution on the bottom of Figure 3-6. The central limit theorem works. We'll revisit the central limit theorem in Chapter 4 when we discuss hypothesis testing.

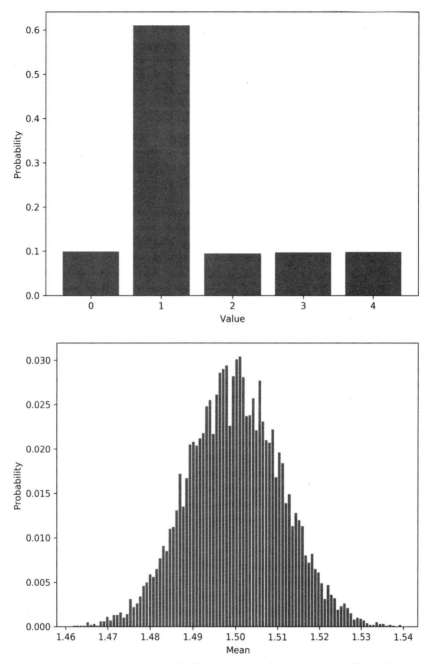

Figure 3-6: An arbitrary discrete distribution (top) and the distribution of sample means drawn from it (bottom)

The Law of Large Numbers

A concept related to the central limit theorem, and often confused with it, is the *law of large numbers*. The law of large numbers states that as the size of a sample from a distribution increases, the mean of the sample moves closer and closer to the mean of the population. In this case, we're contemplating a single sample from the distribution and making a statement about how close we expect its mean to be to the true population mean. For the central limit theorem, we have many different sets of samples from the distribution and are making a statement about the distribution of the means of those sets of samples.

We can demonstrate the law of large numbers quite simply by selecting larger and larger size samples from a distribution and tracking the mean as a function of the sample size (the number of samples drawn). In code, then,

```
m = []
for n in np.linspace(1,8,30):
    t = np.random.normal(1,1,size=int(10**n))
    m.append(t.mean())
```

where we're drawing ever-larger sample sizes from a normal distribution with a mean of 1. The first sample size is 10, and the last is 100 million. If we plot the mean of the samples as a function of sample size, we see the law of large numbers at work.

Figure 3-7 shows the sample means as a function of the number of samples for the normal distribution with a mean of 1 (dashed line). As the number of samples from the distribution increases, the mean of the samples approaches the population mean, which illustrates the law of large numbers.

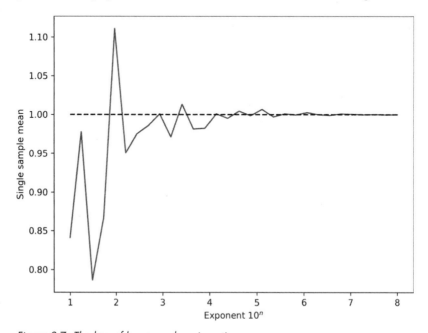

Figure 3-7: The law of large numbers in action

Let's change gears and move on to Bayes' theorem, the last topic for this chapter.

Bayes' Theorem

In Chapter 2, we discussed an example where we determined whether a woman had cancer. There, I promised that Bayes' theorem would tell us how to properly account for the probability of a randomly selected woman in her 40s having breast cancer. Let's fulfill that promise in this section by learning what Bayes' theorem is and how to use it.

Using the product rule, Equation 2.8, we know the following two mathematical statements are true:

$$P(B, A) = P(B|A)P(A)$$

$$P(A, B) = P(A|B)P(B)$$

Additionally, because the joint probability of both A and B doesn't depend on which event we call A and which we call B,

$$P(A, B) = P(B, A)$$

Therefore,

$$P(B|A)P(A) = P(A|B)P(B)$$

Dividing by $P(A)$, we get

$$P(B|A) = \frac{P(A|B)P(B)}{P(A)} \tag{3.1}$$

This is *Bayes' theorem*, the heart of the Bayesian approach to probability, and the proper way to compare two conditional probabilities: $P(B|A)$ and $P(A|B)$. You'll sometimes see Equation 3.1 referred to as *Bayes' rule*. You'll also often see no apostrophe after "Bayes," which is a bit sloppy and ungrammatical, but common.

Equation 3.1 has been enshrined in neon lights, tattoos, and even baby names: "Bayes." The equation is named after Thomas Bayes (1701–1761), an English minister and statistician, and was published after his death. In words, Equation 3.1 says the following:

> The *posterior probability*, $P(B|A)$, is the product of $P(A|B)$, the *likelihood*, and $P(B)$, the *prior*, normalized by $P(A)$, the marginal probability or *evidence*.

Now that we know what Bayes' theorem is, let's see it in action so we can understand it.

Cancer or Not Redux

One way to think about the components of Bayes' theorem is in the context of medical testing. At the beginning of Chapter 2, we calculated the probability of a woman having breast cancer given a positive mammogram and found that it was quite different from what we might naively have believed it to be. Let's revisit that problem now using Bayes' theorem. It might be helpful to reread the first section of Chapter 2 before continuing.

We want to use Bayes' theorem to find the posterior probability, the probability of breast cancer given a positive mammogram. We'll write this as $P(bc + |+)$, meaning breast cancer ($bc+$) given a positive mammogram ($+$).

In the problem, we're told that the mammogram returns a positive result, given the patient has breast cancer, 90 percent of the time. We write this as

$$P(+|bc+) = 0.9$$

This is the likelihood of a positive mammogram in terms of Bayes' equation, $P(A|B) = P(+|bc+)$.

Next, we're told the probability of a random woman having breast cancer is 0.8 percent. Therefore, we know

$$P(bc+) = 0.008$$

This is the prior probability, $P(B)$, in Bayes' theorem.

We have all the components of Equation 3.1 except one: $P(A)$. What is $P(A)$ in this context? It's $P(+)$, the marginal probability of a positive mammogram regardless of any B, any breast cancer status. It's also the evidence that we have, the thing we know: the mammogram was positive.

In the problem, we're told there's a 7 percent chance a woman without breast cancer has a positive mammogram. Is this $P(+)$? No, it is $P(+|bc-)$, the probability of a positive mammogram *given* no breast cancer.

I've referred to $P(A)$ as the marginal probability twice now. We know what to do to get a marginal or total probability: we sum over all the other parts of a joint probability that don't matter for what we want to know. Here, we have to sum over all the partitions of the sample space we don't care about to get the marginal probability of a positive mammogram. What partitions are those? There are only two: either a woman has breast cancer or she doesn't. Therefore, we need to find

$$P(+) = P(+|bc+)P(bc+) + P(+|bc-)P(bc-)$$

We know all of these quantities already, except $P(bc-)$. This is the prior probability that a randomly selected woman will *not* have breast cancer, $P(bc-) = 1 - P(bc+) = 0.992$.

Sometimes, you'll see the summation over other terms in the joint probability expressed in the denominator of Bayes' theorem. Even if they're not explicitly called out, they are there, implicit in what it takes to find $P(A)$.

Finally, we have all the pieces and can calculate the probability using Bayes' theorem:

$$P(bc + \mid +) = \frac{P(+\mid bc+)P(bc+)}{P(+\mid bc+)P(bc+) + P(+\mid bc-)P(bc-)}$$

$$= \frac{0.9(0.008)}{0.9(0.008) + 0.07(0.992)}$$

$$= 0.094 \approx 9 \text{ percent}$$

This is the result we found earlier. Recall, a large percentage of doctors in the study claimed the probability of cancer from a positive mammogram, $P(A\mid B)$, was 90 percent. Their mistake was incorrectly equating $P(A\mid B)$ with $P(B\mid A)$. Bayes' theorem correctly relates the two by using the prior and the marginal probability.

Updating the Prior

We don't need to stop with this single calculation. Consider the following: what if, after a woman receives the news that her mammogram is positive, she decides to have a second mammogram at another facility with different radiologists reading the results, and that mammogram also comes back positive? Does she still believe that her probability of having breast cancer is 9 percent? Intuitively, we might think that she now has more reason to believe that she has cancer. Can this belief be quantified? It can, in the Bayesian view, by updating the prior, $P(bc+)$, with the posterior calculated from the first test, $P(bc + \mid +)$. After all, she now has a stronger prior probability of cancer given the first positive mammogram.

Let's calculate this new posterior based on the previous mammogram result:

$$P(bc + \mid +) = \frac{P(+\mid bc+)P(bc+)}{P(+\mid bc+)P(bc+) + P(+\mid bc-)P(bc-)}$$

$$= \frac{0.9(0.094)}{0.9(0.094) + 0.07(0.906)}$$

$$= 0.572 \approx 57 \text{ percent}$$

As 57 percent is significantly higher than 9 percent, our hypothetical woman now has significantly more reason to believe she has breast cancer.

Notice what has changed in this new calculation, besides a dramatic increase in the posterior probability of breast cancer given the second mammogram's positive result. First, the prior probability of breast cancer went from 0.008 → 0.094, the posterior calculated based on the first test. Second, $P(bc-)$ also changed from 0.992 → 0.906. Why? Because the prior changed and $P(bc-) = 1 - P(bc+)$. The sum of $P(bc+)$ and $P(bc-)$ must still be 1.0— either she has breast cancer, or she doesn't—that's the entire sample space.

In the example above, we updated the prior based on the initial test result, and we had an initial prior given to us in the first example. What about the prior in general? In many cases, Bayesians select the prior, at least initially, based on an actual belief about the problem. Often the prior is a uniform distribution, known as the *uninformed prior* because there's nothing to guide the selection of anything else. For the breast cancer example, the prior is something that can be estimated from an experiment using a random selection of women from the general population.

As mentioned earlier, don't take the numbers here too seriously; they are for example use only. Also, while a woman certainly has the option to get a second opinion, the gold standard for a breast cancer diagnosis is biopsy, the likely next step after an initial positive mammogram. Finally, throughout this section, I've referred to women and breast cancer. Men also get breast cancer, though it is rare, with less than 1 percent of cases in men. However, it made the discussion simpler to refer only to women. I'll note that breast cancer cases in men are more likely to be fatal, though the reasons why are not yet known.

Bayes' Theorem in Machine Learning

Bayes' theorem is prevalent throughout machine learning and deep learning. One classic use of Bayes' theorem, one that can work surprisingly well, is to use it as a classifier. This is known as the *Naive Bayes* classifier. Early email spam filters used this approach quite effectively.

Assume we have a dataset consisting of class labels, y, and feature vectors, x. The goal of a Naive Bayes classifier is to tell us, for each class, the probability that a given feature vector belongs to that class. With those probabilities, we can assign a class label by selecting the largest probability. That is, we want to find $P(y|x)$ for each class label, y. This is a conditional probability, so we can use Bayes' theorem with it:

$$P(y|x) = \frac{P(x|y)P(y)}{P(x)} \tag{3.2}$$

The equation above is saying that the probability of feature vector x representing an instance of class label y is the probability of the class label y generating a feature vector x times the prior probability of class label y occurring, divided by the marginal probability of the feature vector over all class labels. Recall the implicit sum in calculating $P(x)$.

How is this useful to us? Since we have a dataset, we can estimate $P(y)$ using it, assuming the dataset class distribution is a fair representation of what we'd encounter when using the model. And, since we have labels, we can partition the dataset into smaller, per class, collections. This might help us do something useful to get the likelihoods per class, $P(x|y)$. We'll ignore the marginal $P(x)$ completely. Let's see why, in this case, we're free to do so.

Equation 3.2 is for a particular class label, say $y = 1$. We'll have other versions of it for all the class labels in the dataset. We said our classifier consists of calculating the posterior probabilities for each class label and selecting the largest one as the label assigned to an unknown feature vector. The denominator of Equation 3.2 is a scale factor, which makes the output a true probability. For our use case, however, we only care about the relative ordering of $P(y|x)$ over the different class labels. Since $P(x)$ is the same for all y, it's a common factor that will change the number associated with $P(y|x)$ but not the ordering over the different class labels. Therefore, we can ignore it and concentrate on finding the products of the likelihoods and the priors. Although the largest $P(y|x)$ calculated this way is no longer a proper probability, it's still the correct class label to assign.

Given that we can ignore $P(x)$ and the $P(y)$ values are easily estimated from the dataset, we're left with calculating $P(x|y)$, the likelihood that given the class label is y, we'd have a feature vector x. What can we do in this case?

First, we can think about what $P(x|y)$ is. It's a conditional probability for feature vectors given the feature vectors are all representatives of class y. For the moment, let's ignore the y part, since we know the feature vectors all come from class y.

This leaves only $P(x)$ because we fixed y. A feature vector is a collection of individual features, $x = (x_0, x_1, x_2, \ldots, x_{n-1})$ for n features in the vector. Therefore, $P(x)$ is really a joint probability, the probability that all the individual features have their specific values *at the same time*. So, we can write

$$P(x) = P(x_0, x_1, x_2, \ldots, x_{n-1})$$

How does this help? If we make one more assumption about our data, we'll see that we can break up this joint probability in a convenient way. Let's assume that all the features in our feature vector are independent. Recall that *independent* means the value of x_1, say, is in no way affected by the value of any other feature in the vector. This is typically not quite true, and for things like pixels in images *definitely* not true, but we'll assume it's true nonetheless. We're naive to believe it's true, hence the *Naive* in *Naive Bayes*.

If the features are independent, then the probability of a feature taking on any particular value is independent of all the others. In that case, the product rule tells us that we can break the joint probability up like so:

$$P(x) = P(x_0)P(x_1)P(x_2)\ldots P(x_{n-1})$$

This helps tremendously. We have a dataset, labeled by class, allowing us to estimate the probability of any feature for any specific class by counting how often each feature value happens for each class.

Let's put it all together for a hypothetical dataset of three classes—0, 1, and 2—and four features. We first use the dataset, partitioned by class label, to estimate each feature value probability. This provides us the set of $P(x_0), P(x_1)$, and so on, for each feature for each class label. Combined with the prior probability of the class label, estimated from the dataset as the number of each class divided by the total number of samples in the dataset, we calculate for a new unknown feature vector, x,

$$P(0|x) = P(x|0)P(0)$$

$$= P(x_0)P(x_1)P(x_2)P(x_3)P(0)$$

Here, the $P(x_0)$ feature probabilities are specific to class 0 only, and $P(0)$ is the prior probability of class 0 in the dataset. $P(0|x)$ is the unnormalized posterior probability that the unknown feature vector x belongs to class 0. We say *unnormalized* because we're ignoring the denominator of Bayes' theorem, knowing that including it would not change the ordering of the posterior probabilities, only their values.

We can repeat the calculation above to get $P(1|x)$ and $P(2|x)$, making sure to use the per feature probabilities calculated for those classes (the $P(x_0)$s). Finally, we give x the class label for the largest of the three posteriors calculated.

The description above assumes that the feature values are discrete. Usually they aren't, but there are workarounds. One is to bin the feature values to make them discrete. For example, if the feature ranges over $[0, 3]$, create a new feature that is 0, 1, or 2, and assign the continuous feature to one of those bins by truncating any fractional part.

Another workaround is to make one more assumption about the distribution the feature values come from and use that distribution to calculate the $P(x_0)$s per class. Features are often based on measurements in the real world, and many things in the real world follow a normal distribution. Therefore, typically we'd assume that the individual features, while continuous, are normally distributed and we can find estimates of the mean (μ) and standard deviation (σ) from the dataset, per feature, and class label.

Bayes' theorem is useful for calculating probabilities. It's helpful in machine learning as well. The battle between Bayesians and frequentists appears to be waning, though philosophical differences remain. In practice, most researchers are learning that both approaches are valuable, and at times tools from both camps should be used. We'll continue this trend in the next chapter, where we'll examine statistics from a frequentist viewpoint. We defend this decision by pointing out that the vast majority of published scientific results in the last century used statistics this way, which includes the deep learning community, at least when it's presenting the results of experiments.

Summary

This chapter taught us about probability distributions, what they are, and how to draw samples from them, both discrete and continuous. We'll encounter different distributions during our exploration of deep learning. We also discovered Bayes' theorem and saw how it lets us properly relate conditional probabilities. We saw how Bayes' theorem allows us to evaluate the true likelihood of cancer given an imperfect medical test—a common situation. We also learned how to use Bayes' theorem, along with some of the basic probability rules we learned in Chapter 2, to build a simple but often surprisingly effective classifier.

Let's move now into the world of statistics.

4

STATISTICS

Bad datasets lead to bad models. We'd like to understand our data before we build a model, and then use that understanding to create a useful dataset, one that leads to models that do what we expect them to do. Knowing basic statistics will enable us to do just that.

A *statistic* is any number that's calculated from a sample and used to characterize it in some way. In deep learning, when we talk about samples, we're usually talking about datasets. Maybe the most basic statistic is the arithmetic mean, commonly known as the average. The mean of a dataset is a single-number summary of the dataset.

We'll see many different statistics in this chapter. We'll begin by learning about the types of data and characterizing a dataset with summary statistics. Next, we'll learn about quantiles and plotting data to understand what it contains. After that comes a discussion of outliers and missing data. Datasets are seldom perfect, so we need to have some way of detecting bad data and dealing with missing data. We'll follow our discussion of imperfect datasets with a discussion of the correlation between variables. Then we'll close the chapter out by discussing hypothesis testing, where we attempt to answer questions like "How likely is it that the same parent process generated two datasets?" Hypothesis testing is widely used in science, including deep learning.

Types of Data

The four types of data are nominal, ordinal, interval, and ratio. Let's look at each in turn.

Nominal Data

Nominal data, sometimes called *categorical data*, is data that has no ordering between the different values. An example of this is eye color; there is no relationship between brown, blue, and green.

Ordinal Data

For *ordinal data*, the data has a ranking or order, though differences aren't meaningful in a mathematical sense. For example, if a questionnaire asks you to select from "strongly disagree," "disagree," "neutral," "agree," and "strongly agree," it's pretty clear that there is an order. Still, it's also clear that "agree" isn't three more than "strongly disagree." All we can say is that "strongly disagree" is to the left of "agree" (and "neutral" and "disagree").

Another example of ordinal data is education level. If one person has a fourth-grade education and another has an eighth-grade education, we can say that the latter person is more educated than the former, but we can't say that the latter person is twice as educated, because "twice as educated" has no fixed meaning.

Interval Data

Interval data has meaningful differences. For example, if one cup of water is at 40 degrees Fahrenheit and another is at 80 degrees Fahrenheit, we can say that there is a 40-degree difference between the two cups of water. We can't, however, say that there is twice as much heat in the second cup, because the zero for the Fahrenheit scale is arbitrary. Colloquially, we do say it's twice as hot, but in reality, it isn't. To see this, think about what happens if we change the temperature scale to another scale with an arbitrary, though more sensible, zero: the Celsius scale. We see that the first cup is at about 4.4 degrees Celsius, and the second is at 26.7 degrees Celsius. Clearly, the second cup doesn't suddenly now have six times the heat of the first.

Ratio Data

Finally, *ratio data* is data where differences are meaningful, and there is a true zero point. Height is a ratio value because a height of zero is just that—no height at all. Similarly, age is also a ratio value because an age of zero means no age at all. If we were to adopt a new age scale and call a person zero when they reach, say, voting age, we'd then have an interval scale, not a ratio scale.

Let's look at temperature again. We said above that temperature is an interval quantity. This isn't always the case. If we measure temperature in

Fahrenheit or Celsius, then, yes, it is an interval quantity. However, if we measure temperature in Kelvin, the absolute temperature scale, then it becomes a ratio value. Why? Because a temperature of 0 Kelvin (or K) is just that, no temperature at all. If our first cup is at $40°F$, 277.59 K, and the second is at $80°F$, 299.82 K, then we can truthfully say that the second cup is 1.08 times hotter than the first, since $(277.59)(1.08) \approx 299.8$.

Figure 4-1 names the scales and shows their relationships to each other.

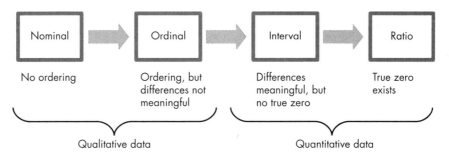

Figure 4-1: The four types of data

Each step in Figure 4-1 from left to right adds something to the data that the type of data on the left is lacking. For nominal to ordinal, we add ordering. For ordinal to interval, we add meaningful differences. Lastly, moving from interval to ratio adds a true zero point.

In practical use, as far as statistics are concerned, we should be aware of the types of data so we don't do something meaningless. If we have a questionnaire, and the mean value of question A on a 1-to-5 rating scale is 2, while for question B it's 4, we can't say that B is rated twice as high as A, only that B was rated higher than A. What "twice" means in this context is unclear and quite probably meaningless.

Interval and ratio data may be continuous (floating-points) or discrete (integers). From a deep learning perspective, models typically treat continuous and discrete data the same way, and we don't need to do anything special for discrete data.

Using Nominal Data in Deep Learning

If we have a nominal value, say a set of colors, such as red, green, and blue, and we want to pass that value into a deep network, we need to change the data before we can use it. As we just saw, nominal data has no order, so while it's tempting to assign a value of 1 to red, 2 to green, and 3 to blue, it would be wrong to do so, since the network will interpret those numbers as interval data. In that case, to the network, blue = 3(red), which is of course nonsense. If we want to use nominal data with a deep network, we need to alter it so that the interval is meaningful. We do this with *one-hot encoding*.

In one-hot encoding, we turn the single nominal variable into a vector, where each element of the vector corresponds to one of the nominal values. For the color example, the one nominal variable becomes a three-element vector with one element representing red, another green, and the last blue. Then, we set the value corresponding to the color to one and all the others to zero, like so:

Value		Vector
red	$\rightarrow$	1 0 0
green	$\rightarrow$	0 1 0
blue	$\rightarrow$	0 0 1

Now the vector values are meaningful because either it's red (1) or it's not (0), green (1) or it's not (0), or blue (1) or it's not (0). The interval between zero and one has mathematical meaning because the presence of the value, say red, is genuinely greater than its absence, and that works the same way for each color. The values are now interval, so the network can use them. In some toolkits, like Keras, class labels are one-hot encoded before passing them to the model. This is done so vector output operates nicely with the one-hot encoded class label when computing the loss function.

Summary Statistics

We're given a dataset. How do we make sense of it? How should we characterize it to understand it better before we use it to build a model?

To answer these questions, we need to learn about *summary statistics*. Calculating summary statistics should be the first thing you do when handed a new dataset. Not looking at your dataset before building a model is like buying a used car without checking the tires, taking it for a test drive, and looking under the hood.

People have different notions of what makes a good set of summary statistics. We'll focus on the following: means; the median; and measures of variation, including variance, standard deviation, and standard error. The range and mode are also often mentioned. The *range* is the difference between the maximum and minimum of the dataset. The *mode* is the most frequent value in the dataset. We generally get a sense of the mode visually from the histogram, as the histogram shows us the shape of the distribution of the data.

Means and Median

Most of us learned how to calculate the average of a set of numbers in elementary school: add the numbers and divide by how many there are. This is the *arithmetic mean*, or, more specifically, the *unweighted* arithmetic mean. If the dataset consists of a set of values, $\{x_0, x_1, x_2, \ldots, x_{n-1}\}$, then the arithmetic mean is the sum of the data divided by the number of elements in the dataset (n). Notationally, we write this as the following.

$$\bar{x} = \frac{1}{n} \sum_{i}^{n-1} x_i \qquad (4.1)$$

The $\bar{x}$ is the typical way to denote the mean of a sample.

Equation 4.1 calculates the unweighted mean. Each value is given a weight of $1/n$, where the sum of all the weights is 1.0. Sometimes, we might want to weight elements of the dataset differently; in other words, not all of them should count equally. In that case, we calculate a weighted mean,

$$\bar{x} = \sum_{i}^{n-1} w_i x_i$$

where w_i is the weight given to x_i and $\sum_i w_i = 1$. The weights are not part of the dataset; they need to come from somewhere else. The grade point average (GPA) used by many universities is an example of a weighted mean. The grade for each course is multiplied by the number of course credits, and the sum is divided by the total number of credits. Algebraically, this is equivalent to multiplying each grade by a weight, $w_i = c_i / \sum_i c_i$, with c_i the number of credits for course i and $\sum_i c_i$ the total number of credits for the semester.

Geometric Mean

The arithmetic mean is by far the most commonly used mean. However, there are others. The *geometric mean* of two positive numbers, a and b, is the square root of their product:

$$\bar{x}_g = \sqrt{ab}$$

In general, the geometric mean of n positive numbers is the nth root of their product:

$$\bar{x}_g = \sqrt[n]{x_0 x_1 x_2 \cdots x_{n-1}}$$

The geometric mean is used in finance to calculate average growth rates. In image processing, the geometric mean can be used as a filter to help reduce image noise. In deep learning, the geometric mean appears in the *Matthews correlation coefficient (MCC)*, one of the metrics we use to evaluate deep learning models. The MCC is the geometric mean of two other metrics, the informedness and the markedness.

Harmonic Mean

The *harmonic mean* of two numbers, a and b, is the reciprocal of the arithmetic mean of their reciprocals:

$$\bar{x}_h = \left(\frac{1}{2} \left(\frac{1}{a} + \frac{1}{b} \right) \right)^{-1}$$

In general,

$$\bar{x}_h = \left(\frac{1}{n} \sum_{i}^{n-1} \frac{1}{x_i}\right)^{-1} = \frac{n}{\frac{1}{x_0} + \frac{1}{x_1} + \cdots + \frac{1}{x_{n-1}}}$$

The harmonic mean shows up in deep learning as the F1 score. This is a frequently used metric for evaluating classifiers. The F1 score is the harmonic mean of the recall (sensitivity) and the precision:

$$F1 = \left(\frac{1}{2}\left(\frac{1}{\text{recall}} + \frac{1}{\text{precision}}\right)\right)^{-1}$$

$$= 2\frac{\text{recall} \cdot \text{precision}}{\text{recall} + \text{precision}}$$

Despite its frequent use, it's not a good idea to use the F1 score to evaluate a deep learning model. To see this, consider the definitions of recall and precision:

$$\text{recall} = \frac{\text{TP}}{\text{TP} + \text{FN}}$$

$$\text{precision} = \frac{\text{TP}}{\text{TP} + \text{FP}}$$

Here, TP is the number of true positives, FN is the number of false negatives, and FP is the number of false positives. These values come from the test set used to evaluate the model. A fourth number that's important for classifiers, TN, is the number of correctly classified true negatives (assuming a binary classifier). The F1 score ignores TN, but to understand how well the model performs, we need to consider both positive and negative classifications. Therefore, the F1 score is misleading and often too optimistic. Better metrics are the MCC mentioned above or Cohen's κ (kappa), which is similar to MCC and usually tracks it closely.

Median

Before moving on to measures of variation, there's one more commonly used summary statistic we'll mention here. It'll show up again a little later in the chapter too. The *median* of a dataset is the middle value. It's the value where, when the dataset is sorted numerically, half the values are below it and half are above it. Let's use this dataset:

$$X = \{55, 63, 65, 37, 74, 71, 73, 87, 69, 44\}$$

If we sort X, we get

$$\{37, 44, 55, 63, 65, 69, 71, 73, 74, 87\}$$

We immediately see a potential problem. I said we need the middle value when the data is sorted. With 10 things in X, there is no middle value. The middle lies between 65 and 69. When the number of elements in the dataset is even, the median is the arithmetic mean of the two middle numbers. Therefore, the median in this case is

$$x_{median} = \frac{65 + 69}{2} = 67$$

The arithmetic mean of the data is 63.8. What's the difference between the mean and the median?

By design, the median tells us the value that splits the dataset, so the number of samples above equals the number below. It's the number of samples that matters. For the mean, it's a sum over the actual data values. Therefore, the mean is sensitive to the values themselves, while the median is sensitive to the ordering of the values.

If we look at X, we see that most values are in the 60s and 70s, with one low value of 37. It's the low value of 37 that drags the mean down relative to the median. An excellent example of this effect is income. The current median annual family income in the United States is about \$62,000. A recent measure of the mean family income in the United States is closer to \$72,000. The difference is because of the small portion of the population who make significantly more money than everyone else. They pull the overall mean up. For income, then, the most meaningful statistic is the median.

Consider Figure 4-2.

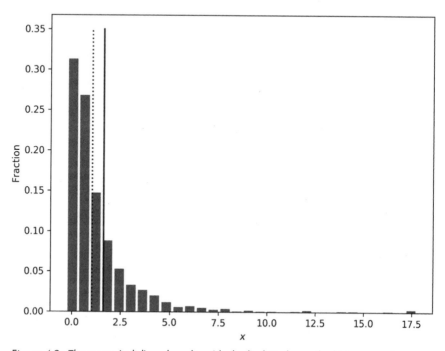

Figure 4-2: The mean (solid) and median (dashed) plotted over the histogram of a sample dataset

Figure 4-2 shows the histogram generated from 1,000 samples of a simulated dataset. Also plotted are the mean (solid line) and median (dashed line). The two do not match; the long tail in the histogram drags the mean up. If we were to count, 500 samples would fall in the bins below the dashed line and 500 in the bins above.

Are there times when the mean and median are the same? Yes. If the data distribution is completely symmetric, then the mean and median will be the same. The classic example of this situation is the normal distribution. Figure 3-4 showed a normal distribution where the left-right symmetry was clear. The normal distribution is special. We'll see it again throughout the chapter. For now, remember that the closer the distribution of the dataset is to a normal distribution, the closer the mean and median will be.

The opposite is also worth remembering: if the dataset's distribution is far from normal, like in Figure 4-2, then the median is likely the better statistic to consider when summarizing the data.

Measures of Variation

A beginning archer shoots 10 arrows at a target. Eight of the beginner's arrows hit the target, two miss completely, and the eight that do hit the target are spread uniformly across it. An expert archer shoots 10 arrows at a target. All of the expert's arrows hit within a few centimeters of the center. Think about the mean position of the arrows. For the expert, all of the arrows are near the center of the target, so we can see that the mean position of the arrows will be near the center. For the beginner, none of the arrows are near the center of the target, but they are scattered more or less equally to the left and right or above and below the center. Because of this, the average position will balance out and be near the center of the target as well.

However, the first archer's arrows are scattered; their location varies greatly. The second archer's arrows, on the other hand, are tightly clustered, and there is little variation in their position. One meaningful way to summarize and understand a dataset is to quantify its variation. Let's see how we might do this.

Deviation vs. Variance

One way we might measure the variation of a dataset is to find the *range*, the difference between the largest and smallest values. However, the range is a crude measurement, as it pays no attention to most of the values in the dataset, only the extremes. We can do better by calculating the mean of the difference between the data values and the mean of the data. The formula is

$$MD = \frac{1}{n} \sum_{i}^{n-1} |x_i - \bar{x}| \tag{4.2}$$

Equation 4.2 is the *mean deviation*. It's a natural measure and gives just what we want: an idea of how far, on average, each sample is from the mean. While there's nothing wrong with calculating the mean deviation, you'll find that it's rarely used in practice. One reason has to do with algebra and calculus. The absolute value is annoying to deal with mathematically.

Instead of the natural measure of variation, let's calculate this one using squared differences:

$$s_n^2 = \frac{1}{n} \sum_{i}^{n-1} (x_i - \bar{x})^2 \qquad (4.3)$$

Equation 4.3 is known as the *biased sample variance*. It's the mean of the squared difference between each value in the dataset and the mean. It's an alternate way of characterizing the scatter in the dataset. Why it's biased, we'll discuss in a second. We'll get into why it's s_n^2 and not s_n shortly after that.

Before we do, it's worth noting that you'll often see a slightly different equation:

$$s^2 = \frac{1}{n-1} \sum_{i}^{n-1} (x_i - \bar{x})^2 \qquad (4.4)$$

This equation is the *unbiased sample variance*. Using $n-1$ in place of n is known as Bessel's correction. It's related to the number of degrees of freedom in the residuals, where the residuals are what's left when the mean is subtracted from each of the values in the dataset. The sum of the residuals is zero, so if there are n values in the dataset, knowing $n-1$ of the residuals allows the last residual to be calculated. This gives us the degrees of freedom for the residuals. We are "free" to calculate $n-1$ of them knowing that we'll get the last one from the fact that the residuals sum to zero. Dividing by $n-1$ gives a less biased estimate of the variance, assuming s_n^2 is biased in some way to begin with.

Why are we talking about biased variance and unbiased variance? Biased how? We should always remember that a dataset is a sample from some parent data-generating process, the population. The true population variance (σ^2) is the scatter of the population around the true population mean (μ). However, we don't know μ or σ^2, so instead, we estimate them from the dataset we do have. The mean of the sample is $\bar{x}$. That's our estimate for μ. It's then natural to calculate the mean of the squared deviations around $\bar{x}$ and call that our estimate for σ^2. That's s_n^2 (Equation 4.3). The claim, which is true but beyond our scope to demonstrate, is that s_n^2 is biased and not the best estimate of σ^2, but if Bessel's correction is applied, we'll have a better estimate of the population variance. So we should use s^2 (Equation 4.4) to characterize the variance of the dataset around the mean.

In summary, we should use $\bar{x}$ and s^2 to quantify the variance of the dataset. Now, why is it s^2? The square root of the variance is the *standard deviation* denoted as σ for the population and s for the estimate of σ calculated from the dataset. Most often, we want to work with the standard deviation. Writing square roots becomes tiresome, so convention has adopted the σ or s notation for the standard deviation and uses the squared form when discussing the variance.

And, because life isn't already ambiguous enough, you'll often see σ used for s, and Equation 4.3 used when it really should be Equation 4.4. Some toolkits, including our beloved NumPy, make it easy to use the wrong formula.

However, as the number of samples in our dataset increases, the difference between the biased and unbiased variance decreases because dividing by n or $n - 1$ matters less and less. A few lines of code illustrate this:

```
>>> import numpy as np
>>> n = 10
>>> a = np.random.random(n)
>>> (1/n)*((a-a.mean())**2).sum()
0.08081748204006689
>>> (1/(n-1))*((a-a.mean())**2).sum()
0.08979720226674098
```

Here, a sample with only 10 values (a) shows a difference in the biased and unbiased variance in the third decimal. If we increase our dataset size from 10 to 10,000, we get

```
>>> n = 10000
>>> a = np.random.random(n)
>>> (1/n)*((a-a.mean())**2).sum()
0.08304350577482553
>>> (1/(n-1))*((a-a.mean())**2).sum()
0.08305181095592111
```

The difference between the biased and unbiased estimate of the variance is now in the fifth decimal. Therefore, for the large datasets we typically work with in deep learning, it matters little in practice whether we use s_n or s for the standard deviation.

MEDIAN ABSOLUTE DEVIATION

The standard deviation is based on the mean. The mean, as we saw above, is sensitive to extreme values, and the standard deviation is doubly so because we square the deviation from the mean for each sample. A measure of variability that is insensitive to extreme values in the dataset is the *median absolute deviation (MAD)*. The MAD is defined as the median of the absolute values of the difference between the data and the median:

$$MAD = \text{median}(|X_i - \text{median}(X)|)$$

Procedurally, first calculate the median of the data, then subtract it from each data value, making the result positive, and report the median of that set. The implementation is straightforward:

```
def MAD(x):
    return np.median(np.abs(x-np.median(x)))
```

The MAD is not often used, but its insensitivity to extreme values in the dataset argues toward more frequent use, especially for outlier detection.

Standard Error vs. Standard Deviation

We have one more measure of variance to discuss: the *standard error of the mean (SEM)*. The SEM is often simply called the *standard error (SE)*. We need to go back to the population to understand what the SE is and when to use it. If we select a sample from the population, a dataset, we can calculate the mean of the sample, $\bar{x}$. If we choose repeated samples and calculate those sample means, we'll generate a dataset of means of the samples from the population. This might sound familiar; it's the process we used to illustrate the central limit theorem in Chapter 3. The standard deviation of the set of means is the standard error.

The formula for the standard error from the standard deviation is straightforward,

$$SE = \frac{s}{\sqrt{n}}$$

and is nothing more than a scaling of the sample standard deviation by the square root of the number of samples.

When should we use the standard deviation, and when should we use the standard error? Use the standard deviation to learn about the distribution of the samples around the mean. Use the standard error to say something about how good an estimate of the population mean a sample mean is. In a sense, the standard error is related to both the central limit theorem, as that affects the standard deviation of the means of multiple samples from the parent population, and the law of large numbers, since a larger dataset is more likely to give a better estimate of the population mean.

From a deep learning point of view, we might use the standard deviation to describe the dataset used to train a model. If we train and test several models, remembering the stochastic nature of deep network initialization, we can calculate a mean over the models for some metric, say the accuracy. In that case, we might want to report the mean accuracy plus or minus the standard error. As we train more models and gain confidence that the mean accuracy represents the sort of accuracy the model architecture can provide, we should expect that the error in the mean accuracy over the models will decrease.

To recap, in this section, we discussed different summary statistics, values we can use to start to understand a dataset. These include the various means (arithmetic, geometric, and harmonic), the median, the standard deviation, and, when appropriate, the standard error. For now, let's see how we can use plots to help understand a dataset.

Quantiles and Box Plots

To calculate the median, we need to find the middle value, the number splitting the dataset into two halves. Mathematically, we say that the median divides the dataset into two quantiles.

A *quantile* splits the dataset into fixed-sized groups where the fixed size is the number of data values in the quantile. Since the median splits the dataset into two equally sized groups, it's a *2-quantile*. Sometimes you'll see the median referred to as the *50th percentile*, meaning 50 percent of the data values are less than this value. By similar reasoning, then, the 95th percentile is the value that 95 percent of the dataset is less than. Researchers often calculate 4-quantiles and refer to them as *quartiles*, since they split the dataset into four groups such that 25 percent of the data values are in the first quartile, 50 percent are in the first and second, and 75 percent are in the first, second, and third, with the final 25 percent in the fourth quartile.

Let's work through an example to understand what we mean by quantiles. The example uses a synthetic exam dataset representing 1,000 test scores. See the file *exams.npy*. We'll use NumPy to calculate the quartile values for us and then plot a histogram of the dataset with the quartile values marked. First, let's calculate the quartile positions:

```
d = np.load("exams.npy")
p = d[:,0].astype("uint32")
q = np.quantile(p, [0.0, 0.25, 0.5, 0.75, 1.0])

print("Quartiles: ", q)
print("Counts by quartile:")
print("    %d" % ((q[0] <= p) & (p < q[1])).sum())
print("    %d" % ((q[1] <= p) & (p < q[2])).sum())
print("    %d" % ((q[2] <= p) & (p < q[3])).sum())
print("    %d" % ((q[3] <= p) & (p < q[4])).sum())
```

This code, along with code to generate the plot, is in the file *quantiles.py*.

First we load the synthetic exam data and keep the first exam scores (p). Note, we make p an integer array so we can use np.bincount later to make the histogram. (That code is not shown above.) We then use NumPy's np.quantile function to calculate the quartile values. This function takes the source array and an array of quantile values in the range $[0, 1]$. The values are fractions of the distance from the minimum value of the array to its maximum. So, asking for the 0.5 quantile is asking for the value that is half the distance between the minimum of p and its maximum such that the number of values in each set is equal.

To get quartiles, we ask for the 0.25, 0.5, and 0.75 quantiles to get the values such that 25 percent, 50 percent, and 75 percent of the elements of p are less than the values. We also ask for the 0.0 and 1.0 quantiles, the minimum and maximum of p. We do this for convenience when we count the number of elements in each range. Note, we could have instead used the np.percentile function. It returns the same values as np.quantile but uses percentage values instead of fractions. In that case, the second argument would have been [0,25,50,75,100].

The returned quartile values are in q. We print them to get

```
18.0, 56.75, 68.0, 78.0, 100.0
```

Here, 18 is the minimum, 100 is the maximum, and the three cutoff values for the quartiles are 56.75, 68, and 78. Note that the cutoff for the second quartile is the median, 68.

The remaining code counts the number of values in p in each range. With 1,000 values, we'd expect to have 250 in each range, but because the math doesn't always fall along existing data values, we get instead

```
250, 237, 253, 248
```

meaning 250 elements of p are less than 56.75, 237 are in $[56.75, 68]$, and so forth.

The code above uses a clever counting trick worth explaining. We want to count the number of values in p in some range. We can't use NumPy's np.where function, as it doesn't like the compound conditional statement. However, if we use an expression like 10 <= p, we'll be given an array the same size as p where each element is either True if the condition is true for that element or False if it is not. Therefore, asking for 10 <= p and p < 90 will return two Boolean arrays. To get the elements where both conditions are true, we need to logically AND them together (&). This gives us a final array the same size and shape as p, where all True elements represent values in p in $[10, 90)$. To get the count, we apply the sum method that for a Boolean array treats True as one and False as zero.

Figure 4-3 shows the histogram of the exam data with the quartiles marked.

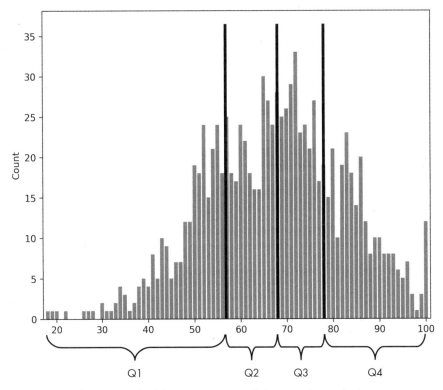

Figure 4-3: A histogram of 1,000 exam scores with the quartiles marked

The example above shows yet again how useful a histogram is for visualizing and understanding data. We should use histograms whenever possible to help understand what's going on with a dataset. Figure 4-3 superimposes the quartile values on the histogram. This helps us understand what the quartiles are and their relationship to the data values, but this is not a typical presentation style. More typical, and useful because it can show multiple features of a dataset, is the *box plot*. Let's use it now for the exam scores above, but this time we'll also include the two other sets of exam scores we ignored previously.

We'll show a box plot first, and then explain it. To see the box plot for the three exams in the *exams.npy* file, use

```
d = np.load("exams.npy")
plt.boxplot(d)
plt.xlabel("Test")
plt.ylabel("Scores")
plt.show()
```

where we're loading the full set of exam scores and then using the Matplotlib boxplot function.

Take a look at the output, shown in Figure 4-4.

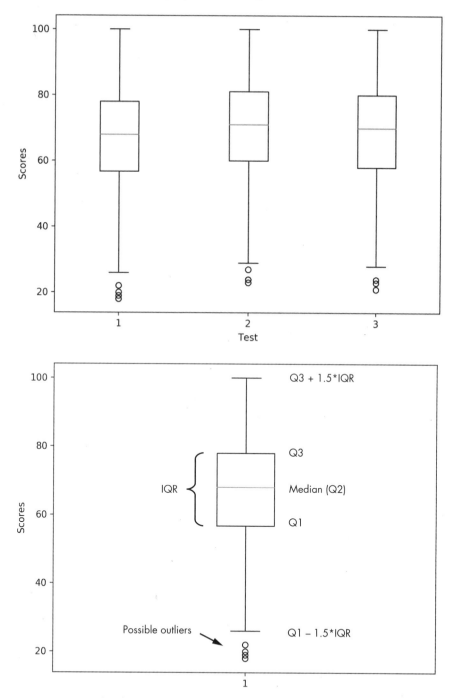

Figure 4-4: Box plots for the three exams (top), and the box plot for the first exam with the components marked (bottom)

The top chart in Figure 4-4 shows the box plot for the three sets of exam scores in |exams.npy|. The first of these is plotted again on the bottom of Figure 4-4, along with labels describing the parts of the plot.

A box plot shows us a visual summary of the data. The box in the bottom chart in Figure 4-4 illustrates the range between the cutoffs for the first quartile (Q1) and the third quartile (Q3). The numerical difference between Q3 and Q1 is known as the *interquartile range (IQR)*. The larger the IQR, the more spread out the data is around the median. Notice that the score is on the y-axis this time. We could have easily made the plot horizontal, but vertical is the default. The median (Q2) is marked near the middle of the box. The mean is not shown in a box plot.

The box plot includes two additional lines, the *whiskers*, though Matplotlib calls them *fliers*. As indicated, they are 1.5 times the IQR above Q3 or below Q1. Finally, there are some circles labeled "possible outliers." By convention, values outside of the whiskers are considered *possible outliers*, meaning they might represent erroneous data, either entered incorrectly by hand or, more likely these days, received from faulty sensors. For example, bright or dead pixels on a CCD camera might be considered outliers. When evaluating a potential dataset, we should be sensitive to outliers and use our best judgment about what to do with them. Usually, there are only a few, and we can drop the samples from the dataset without harm. However, it's also possible that the outliers are actually real and are highly indicative of a particular class. If that's the case, we want to keep them in the dataset in the hopes that the model will use them effectively. Experience, intuition, and common sense must guide us here.

Let's interpret the top chart in Figure 4-4 showing the three sets of exam scores. The top of the whiskers is at 100 each time, which makes sense: a 100 is a perfect score, and there were 100s in the dataset. Notice that the box portion of the plot is not centered vertically in the whiskers. Recalling that 50 percent of the data values are between Q1 and Q3, with 25 percent above and below Q2 in the box, we see that the data is not rigorously normal; its distribution deviates from a normal curve. A glance back to the histogram in Figure 4-3 confirms this for the first exam. Similarly, we see the second and third exams deviate from normality as well. So, a box plot can tell us how similar the distribution of the dataset is to a normal distribution. When we discuss hypothesis testing below, we'll want to know if the data is normally distributed or not.

What about possible outliers, the values below $Q1 - 1.5 \times IQR$? We know the dataset represents test scores, so common sense tells us that these are not outliers but valid scores by particularly confused (or lazy) students. If the dataset contained values above 100 or below zero, those would be fair game to label outliers.

Sometimes dropping samples with outliers is the right thing to do. However, if the outlier is caused by missing data, cutting the sample might not be an option. Let's take a look at what we might do with missing data, and why we should generally avoid it like the plague.

Missing Data

Missing data is just that, data we don't have. If the dataset consists of samples representing feature vectors, missing data shows up as one or more features in a sample that were not measured for some reason. Often, missing data is encoded in some way. If the value is only positive, a missing feature might be marked with a −1 or, historically, −999. If the feature is given to us as a string, the string might be empty. For floating-point values, a not a number (NaN) might be used. NumPy makes it easy for us to check for NaNs in an array by using `np.isnan`:

```
>>> a = np.arange(10, dtype="float64")
>>> a[3] = np.nan
>>> np.isnan(a[3])
True
>>> a[3] == np.nan
False
>>> a[3] is np.nan
False
```

Notice that direct comparison to `np.nan` with either `==` or `is` doesn't work; only testing with `np.isnan` works.

Detecting missing data is dataset-specific. Assuming we've convinced ourselves there is missing data, how do we handle it?

Let's generate a small dataset with missing values and use our existing statistics knowledge to see how to handle them. The code for the following is in `missing.py`. First, we generate a dataset of 1,000 samples, each with four features:

```
N = 1000
np.random.seed(73939133)
x = np.zeros((N,4))
x[:,0] = 5*np.random.random(N)
x[:,1] = np.random.normal(10,1,size=N)
x[:,2] = 3*np.random.beta(5,2,N)
x[:,3] = 0.3*np.random.lognormal(size=N)
```

The dataset is in x. We fix the random number seed to get a reproducible result. The first feature is uniformly distributed. The second is normally distributed, while the third follows a beta distribution and the fourth a lognormal distribution.

At the moment, x has no missing values. Let's add some by making random elements NaNs:

```
i = np.random.randint(0,N, size=int(0.05*N))
x[i,0] = np.nan
i = np.random.randint(0,N, size=int(0.05*N))
x[i,1] = np.nan
i = np.random.randint(0,N, size=int(0.05*N))
```

```
x[i,2] = np.nan
i = np.random.randint(0,N, size=int(0.05*N))
x[i,3] = np.nan
```

The dataset now has NaNs across 5 percent of its values.

If a few samples in a large dataset have missing data, we can remove them from the dataset with little worry. However, if 5 percent of the samples have missing data, we probably don't want to lose that much data. More worrisome still, what if there's a correlation between the missing data and a particular class? Throwing the samples away might bias the dataset in some way that'll make the model less useful.

So, what can we do? We just spent many pages learning how to summarize a dataset with basic descriptive statistics. Can we use those? Of course. We can look at the distributions of the features, ignoring the missing values, and use those distributions to decide how we might want to replace the missing data. Naively, we'd use the mean of the data we do have, but looking at the distribution may or may not push us toward the median instead, depending on how far the distribution is from normal. This sounds like a job for a box plot. Fortunately for us, Matplotlib's `boxplot` function is smart; it ignores the NaNs. Therefore, making the box plot is a straightforward call to `boxplot(x)`.

Figure 4-5 shows us the dataset with the NaNs ignored.

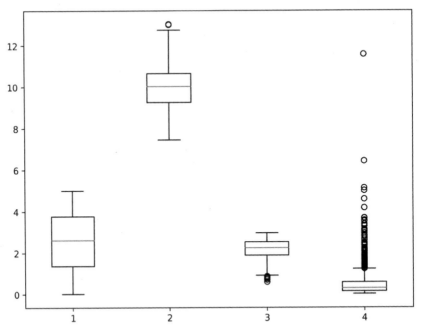

Figure 4-5: Box plot of the dataset ignoring missing values

The boxes in Figure 4-5 make sense for the distributions of the features. Feature 1 is uniformly distributed, so we expect a symmetric box around the mean/median. (These are the same for the uniform distribution.) Feature 2 is normally distributed, so we get a similar box structure as Feature 1, but, with only 1,000 samples, some asymmetry is evident. The beta distribution of Feature 3 is skewed toward the top of its range, which we see in the box plot. Finally, the lognormal distribution of Feature 4 should be skewed toward lower values, with a long tail visible as the many "outliers" above the whiskers, an object lesson against mindlessly calling such values outliers.

Because we have features that are highly not normally distributed, we'll update missing values with the median instead of the mean. The code is straightforward:

```
good_idx = np.where(np.isnan(x[:,0]) == False)
m = np.median(x[good_idx,0])
bad_idx = np.where(np.isnan(x[:,0]) == True)
x[bad_idx,0] = m
```

Here, i first holds the indices of Feature 1 that are not NaNs. We use these to calculate the median (m). Next, we set i to the indices that are NaNs and replace them with the median. We can do the same for the other features, updating the entire dataset so we no longer have missing values.

Did we cause much of a change from the earlier distributions? No, because we only updated 5 percent of the values. For example, for Feature 3, based on the beta distribution, the mean and standard deviations change like so:

```
non-NaN mean, std =  2.169986, 0.474514
updated mean, std =  2.173269, 0.462957
```

The moral of the story is that if there's enough missing data that the dataset might become biased by dropping it, the safest thing to do is replace the missing data with the mean or median. To decide whether to use the mean or median, consult descriptive statistics, a box plot, or a histogram.

Additionally, if the dataset is labeled, as a deep learning dataset would be, the process described above needs to be completed with the mean or median of samples grouped by each class. Otherwise, the calculated value might be inappropriate for the class.

With missing data eliminated, deep learning models can be trained on the dataset.

Correlation

At times, there is an association between the features in a dataset. If one goes up, the other might go up as well, though not necessarily in a simple linear way. Or, the other might go down—a negative association. The proper word for this type of association is *correlation*. A statistic that measures correlation is a handy way to understand how the features in a dataset are related.

For example, it isn't hard to see that the pixels of most images are highly correlated. This means if we select a pixel at random and then an adjacent pixel, there's a good chance the second pixel will be similar to the first pixel. Images where this is not true look to us like random noise.

In traditional machine learning, highly correlated features were undesirable, as they didn't add any new information and only served to confuse the models. The entire art of feature selection was developed, in part, to remove this effect. For modern deep learning, where the network itself learns a new representation of the input data, it's less critical to have uncorrelated inputs. This is, in part, why images work as inputs to deep networks when they usually fail to work at all with older machine learning models.

Whether the learning is traditional or modern, as part of summarizing and exploring a dataset, correlations among the features are worth examining and understanding. In this section, we'll discuss two types of correlations. Each type returns a single number that measures the strength of the correlation between two features in the dataset.

Pearson Correlation

The *Pearson correlation coefficient* returns a number, $r \in [-1, +1]$, that indicates the strength of the *linear* correlation between two features. By *linear* we mean how strongly we can describe the correlation between the features by a line. If the correlation is such that one feature goes up exactly as the other feature goes up, the correlation coefficient is +1. Conversely, if the second feature goes down exactly as the other goes up, the correlation is −1. A correlation of zero means there is no association between the two features; they are (possibly) independent.

I slipped the word *possibly* in the sentence above because there are situations where a nonlinear dependence between two features might lead to a zero Pearson correlation coefficient. These situations are not common, however, and for our purposes, we can claim a correlation coefficient near zero indicates the two features are independent. The closer the correlation coefficient is to zero, either positive or negative, the weaker the correlation between the features.

The Pearson correlation is defined using the means of the two features or the means of products of the two features. The inputs are two features, two columns of the dataset. We'll call these inputs X and Y, where the capital letter refers to a vector of data values. Note, since these are two features from the dataset, X_i is paired with Y_i, meaning they both come from the same feature vector.

The formula for the Pearson correlation coefficient is

$$\text{corr}(X, Y) = \frac{\text{E}(XY) - \text{E}(X)\text{E}(Y)}{\sqrt{\text{E}(X^2) - \text{E}(X)^2}\sqrt{\text{E}(Y^2) - \text{E}(Y)^2}} \qquad (4.5)$$

We've introduced a new, but commonly used, notation. The mean of X is the *expectation* of X, denoted as $E(X)$. Therefore, in Equation 4.5, we see the mean of X, $E(X)$, and the mean of Y, $E(Y)$. As we might suspect, $E(XY)$ is the mean of the product of X and Y, element by element. Similarly, $E(X^2)$ is the mean of the product of X with itself, and $E(X)^2$ is the square of the mean of X. With this notation in hand, we can easily write our own function to calculate the Pearson correlation of two vectors of features:

```
import numpy as np
def pearson(x,y):
    exy = (x*y).mean()
    ex = x.mean()
    ey = y.mean()
    exx = (x*x).mean()
    ex2 = x.mean()**2
    eyy = (y*y).mean()
    ey2 = y.mean()**2
    return (exy - ex*ey)/(np.sqrt(exx-ex2)*np.sqrt(eyy-ey2))
```

The pearson function directly implements Equation 4.5.

Let's set up a scenario where we can use pearson and compare it to what NumPy and SciPy provide. The code that follows, including the definition of pearson above, is in the file *correlation.py*.

First, we'll create three correlated vectors, x, y, and z. We imagine that these are features from a dataset so that x[0] is paired with y[0] and z[0]. The code we need is

```
np.random.seed(8675309)
N = 100
x = np.linspace(0,1,N) + (np.random.random(N)-0.5)
y = np.random.random(N)*x
z = -0.1*np.random.random(N)*x
```

Notice that we're again fixing the NumPy pseudorandom seed to make the output reproducible. The first feature, x, is a noisy line from zero to one. The second, y, tracks x but is also noisy because of the multiplication by a random value in $[0, 1]$. Finally, z is negatively correlated to x because of the -0.1 coefficient.

The top chart in Figure 4-6 plots the three feature values sequentially to see how they track each other. The bottom chart shows the three as paired points, with one value on the x-axis and the other on the y-axis.

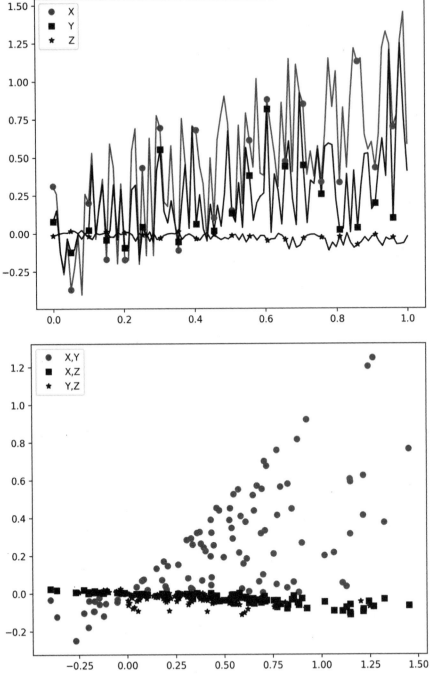

Figure 4-6: Three features in sequence to show how they track (top), and a scatter plot of the features as pairs (bottom)

The NumPy function to calculate the Pearson correlation is `np.corrcoef`. Unlike our version, this function returns a matrix showing the correlations between all pairs of variables passed to it. For example, using our pearson function, we get the following as the correlation coefficients between x, y, and z:

```
pearson(x,y):  0.682852
pearson(x,z): -0.850475
pearson(y,z): -0.565361
```

NumPy returns the following, with x, y, and z stacked as a single 3×100 array:

```
>>> d = np.vstack((x,y,z))
>>> print(np.corrcoef(d))
[[ 1.          0.68285166 -0.85047468]
 [ 0.68285166  1.         -0.56536104]
 [-0.85047468 -0.56536104  1.         ]]
```

The diagonal corresponds to the correlation with each feature and itself, which is naturally perfect and therefore 1.0. The correlation between x and y is in element 0,1 and matches our pearson function value. Similarly, the correlation between x and z is in element 0,2, and the correlation between y and z is in element 1,2. Notice also that the matrix is symmetric, which we expect because $corr(X, Y) = corr(Y, X)$.

SciPy's correlation function is stats.pearsonr, which acts like ours but returns a *p*-value along with the *r* value. We'll discuss *p*-values more later in the chapter. We use the returned *p*-value as the probability of an uncorrelated system producing the calculated correlation value. For our example features, the *p*-value is virtually identical to zero, implying there's no reasonable likelihood that an uncorrelated system produced the features.

We stated earlier that for images, nearby pixels are usually highly correlated. Let's see if this is actually true for a sample image. We'll use the China image included with sklearn and treat specific rows of the green band as the paired vectors. We'll calculate the correlation coefficient for two adjacent rows, a row further away, and a random vector:

```
>>> from sklearn.datasets import load_sample_image
>>> china = load_sample_image('china.jpg')
>>> a = china[230,:,1].astype("float64")
>>> b = china[231,:,1].astype("float64")
>>> c = china[400,:,1].astype("float64")
>>> d = np.random.random(640)
>>> pearson(a,b)
0.8979360
>>> pearson(a,c)
-0.276082
>>> pearson(a,d)
-0.038199
```

Comparing row 230 and row 231 shows that they are highly positively correlated. Comparing rows 230 and 400 shows a weaker and, in this case, negative correlation. Finally, as we might expect, correlation with a random vector gives a value approaching zero.

The Pearson correlation coefficient is so widely used that you'll often see it referred to as merely *the correlation coefficient*. Let's now take a look at a second correlation function and see how it differs from the Pearson coefficient.

Spearman Correlation

The second correlation measure we'll explore is the *Spearman correlation coefficient*, $\rho \in [-1, +1]$. It's a measure based on the ranks of the feature values instead of the values themselves.

To rank X, we replace each value in X with the index to that value in the sorted version of X. If X is

[86, 62, 28, 43, 3, 92, 38, 87, 74, 11]

then the ranks are

[7, 5, 2, 4, 0, 9, 3, 8, 6, 1]

because when X is sorted, 86 goes in the eighth place (counting from zero), and 3 goes first.

The Pearson correlation looks for a linear relationship, whereas the Spearman looks for any monotonic association between the inputs.

If we have the ranks for the feature values, then the Spearman coefficient is

$$\rho = 1 - \left(\frac{6}{n(n^2 - 1)} \right) \sum_{i=0}^{n-1} d_i^2 \tag{4.6}$$

where n is the number of samples and $d = \text{rank}(X) - \text{rank}(Y)$ is the difference of the rank of the paired X and Y values. Note how Equation 4.6 is only valid if the rankings are unique (that is, there are no repeated values in X or Y).

To calculate d in Equation 4.6, we need to rank X and Y and use the difference of the ranks. The Spearman correlation is the Pearson correlation of the ranks.

The example above points the way to an implementation of the Spearman correlation:

```
import numpy as np
def spearman(x,y):
    n = len(x)
    t = x[np.argsort(x)]
    rx = []
    for i in range(n):
```

```
        rx.append(np.where(x[i] == t)[0][0])
    rx = np.array(rx, dtype="float64")
    t = y[np.argsort(y)]
    ry = []
    for i in range(n):
        ry.append(np.where(y[i] == t)[0][0])
    ry = np.array(ry, dtype="float64")
    d = rx - ry
    return 1.0 - (6.0/(n*(n*n-1)))*(d**2).sum()
```

To get the ranks, we need to first sort X (t). Then, for each value in X (x), we find where it occurs in t via np.where and take the first element, the first match. After building the rx list, we make it a floating-point NumPy array. We do the same for Y to get ry. With the ranks, d is set to their difference, and Equation 4.6 is used to return the Spearman ρ value.

Please note that this version of the Spearman correlation is limited by Equation 4.6 and should be used when there are no duplicate values in X or Y. Our example in this section uses random floating-point values, so the probability of an exact duplicate is quite low.

We'll compare our spearman implementation to the SciPy version, stats .spearmanr. Like the SciPy version of the Pearson correlation, stats.spearmanr returns a p-value. We'll ignore it. Let's see how our function compares:

```
>>> from scipy.stats import spearmanr
>>> print(spearman(x,y), spearmanr(x,y)[0])
0.694017401740174 0.6940174017401739
>>> print(spearman(x,z), spearmanr(x,z)[0])
-0.8950855085508551 -0.895085508550855
>>> print(spearman(y,z), spearmanr(y,z)[0])
-0.6414041404140414 -0.6414041404140414
```

We have complete agreement with the SciPy function out to the last bit or so of the floating-point value.

It's important to remember the fundamental difference between the Pearson and Spearman correlations. For example, consider the correlation between a linear ramp and the sigmoid function:

```
ramp = np.linspace(-20,20,1000)
sig = 1.0 / (1.0 + np.exp(-ramp))
print(pearson(ramp,sig))
print(spearman(ramp,sig))
```

Here, ramp increases linearly from −20 to 20 and sig follows a sigmoid shape ("S" curve). The Pearson correlation will be on the high side, since both are increasing as x becomes more positive, but the association is not purely linear. Running the example gives

```
0.905328
1.0
```

indicating a Pearson correlation of 0.9 but a perfect Spearman correlation of 1.0, since for every increase in ramp there is an increase in sig and *only* an increase. The Spearman correlation has captured the nonlinear relationship between the arguments, while the Pearson correlation has only hinted at it. If we're analyzing a dataset intended for a classical machine learning algorithm, the Spearman correlation might help us decide which features to keep and which to discard.

This concludes our examination of statistics for describing and understanding data. Let's now learn how to use hypothesis testing to interpret experimental results and answer questions like "Are these two sets of data samples from the same parent distribution?"

Hypothesis Testing

We have two independent sets of 50 students studying cell biology. We have no reason to believe the groups differ in any significant way, as students from the larger population were assigned randomly. Group 1 attended the lectures and, in addition, worked through a structured set of computer exercises. Group 2 only attended the lectures. Both groups took the same final examination, leading to the test scores given in Table 4-1. We want to know if asking the students to work through the computer exercises made a difference in their final test scores.

Table 4-1: Group 1 and Group 2 Test Scores

Group 1	81 80 85 87 83 87 87 90 79 83 88 75 87 92 78 80
	83 91 82 88 89 92 97 82 79 82 82 85 89 91 83 85
	77 81 90 87 82 84 86 79 84 85 90 84 90 85 85 78
	94 100
Group 2	92 82 78 74 86 69 83 67 85 82 81 91 79 82 82 88
	80 63 85 86 77 94 85 75 77 89 86 71 82 82 80 88
	72 91 90 92 95 87 71 83 94 90 78 60 76 88 91 83
	85 73

Figure 4-7 shows a box plot of Table 4-1.

To understand if there is a significant change in final test scores between the two groups, we need to test some hypotheses. The method we'll use to test the hypotheses is known as *hypothesis testing*, and it's a critical piece of modern science.

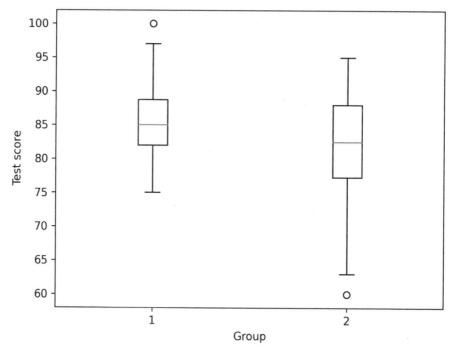

Figure 4-7: Box plot for the data in Table 4-1

Hypothesis testing is a broad topic, too extensive for us to provide more than a minimal introduction here. As this is a book on deep learning, we'll focus on the scenario a deep learning researcher is likely to encounter. We'll consider only two hypothesis tests: the t-test for unpaired samples of differing variance (a parametric test) and the Mann-Whitney U (a nonparametric test). As we progress, we'll understand what these tests are and why we're restricting ourselves to them, as well as the meaning of *parametric* and *nonparametric*.

To be successful with hypothesis testing, we need to know what we mean by *hypothesis*, so we'll address that first, along with our rationale for limiting the types of hypothesis testing we'll consider. With the hypothesis concept in hand, we'll discuss the t-test and the Mann-Whitney U test in turn, using the data in Table 4-1 as our example. Let's get started.

Hypotheses

To understand if two sets of data are from the same parent distribution or not, we might look at summary statistics. Figure 4-7 shows us the box plot for Group 1 and Group 2. It appears that the two groups have different means and standard deviations. How do we know? The box plot shows us the location of the medians, and the whiskers tell us something about the variance. Both of these together hint that the means will be different because the medians are different, and both sets of data are reasonably symmetric around the median. The space between the whiskers hints at the standard deviation. So, let's make hypotheses using the means of the datasets.

In hypothesis testing, we have two hypotheses. The first, known as the *null hypothesis* (H_0), is that the two sets of data *are* from the same parent distribution, that there is nothing special to differentiate them. The second hypothesis, the *alternative hypothesis* (H_a), is that the two groups are not from the same distribution. Since we'll be using the means, H_0 is saying that the means, really the means of the parent population that generated the data, are the same. Similarly, if we reject H_0, we are implicitly accepting H_a and claiming we have evidence that the means are different. We don't have the true population means, so we'll use the sample means and standard deviations instead.

Hypothesis testing doesn't tell us definitively whether H_0 is true. Instead, it gives us evidence in favor of rejecting or accepting the null hypothesis. It's critical to remember this.

We're testing two independent samples to see if we should think of them as coming from the same parent distribution. There are other ways to use hypothesis testing, but we rarely encounter them in deep learning. For the task at hand, we need the sample means and the sample standard deviations. Our tests will ask the question, "Is there a meaningful difference in the means of these two sets?"

We're only interested in detecting whether the two groups of data are from the same parent distribution, so another simplification we'll make is that all of our tests will be *two-sided*, or *two-tailed*. When we use a test, like the t-test we'll describe next, we're comparing our calculated test statistic (the t-value) to the distribution of the test statistic and asking questions about how likely our calculated t-value is. If we want to know about the test statistic being above or below some fraction of that distribution, we're making a two-sided test. If instead we want to know about the likelihood of the test statistic being above a particular value without caring about it being below, or vice versa, then we're making a one-sided test.

Let's lay out our assumptions and approach:

1. We have two independent sets of data we wish to compare.

2. We're making no assumption as to whether the standard deviations of the data are the same.

3. Our null hypothesis is that the means of the parent distributions of the datasets are the same, $H_0 : \mu_1 = \mu_2$. We'll use the sample means ($\bar{x}_1, \bar{x}_2$) and sample standard deviations (s_1, s_2) to help us decide to accept or reject H_0.

4. Hypothesis tests assume that the data is *independent and identically distributed (i.i.d.)*. We interpret this as a statement that the data is a fair random sample.

With these assumptions understood, let's start with the t-test, the most widely used hypothesis test.

The t-test

The *t-test* depends on *t*, the test statistic. This statistic is compared to the t-distribution and used to generate a *p*-value, a probability we'll use to reach a conclusion about H_0. There's a rich history behind the t-test and the related z-test that we'll ignore here. I encourage you to dive more deeply into hypothesis testing when you have the chance or, at a minimum, review thoughtful articles about the proper way to do a hypothesis test and interpret its results.

The t-test is a *parametric* test. This means there are assumptions about the data and the distribution of the data. Specifically, the t-test assumes, beyond the data being i.i.d., that the distribution (histogram) of the data is normal. We've stated before that many physical processes do seem to follow a normal distribution, so there's reason to think that data from actual measurements might do so.

There are many ways to test if a dataset is normally distributed, but we'll ignore them, as there's some debate about the utility of such tests. Instead, I'll (somewhat recklessly) suggest you use the t-test and the Mann-Whitney U test together to help make your decision about accepting or rejecting H_0. Using both tests might lead to a situation where they disagree, where one test says there's evidence against the null hypothesis and the other says there isn't. In general, if the nonparametric test is claiming evidence against H_0, then one should probably accept that evidence regardless of the t-test result. If the t-test result is against H_0, but the Mann-Whitney U test isn't, and you think the data is normal, then you might also accept the t-test result.

The t-test has different versions. We explicitly stated above that we'll use a version designed for datasets of differing size and variance. The specific version of the t-test we'll use is *Welch's t-test*, which doesn't assume the variance of the two datasets is the same.

The t-score for Welch's t-test is

$$t = \frac{\bar{x}_1 - \bar{x}_2}{\sqrt{\frac{s_1^2}{n_1} + \frac{s_2^2}{n_2}}}$$

where n_1 and n_2 are the size of the two groups.

The t-score, and an associated value known as the *degrees of freedom*, which is similar to but also different from the degrees of freedom mentioned above, generates the appropriate t-distribution curve. To get a *p*-value, we calculate the area under the curve, both above and below (positive and negative t-score), and return it. Since the integral of a probability distribution is 1, the total area under the tails from the positive and negative t-score value to positive and negative infinity will be the *p*-value. We'll use the degrees of freedom below to help us calculate confidence intervals.

What does the *p*-value tell us? It tells us the probability of seeing the difference between the two means we see, or larger, *if* the null hypothesis is true. Typically, if this probability is below some threshold we've chosen, we reject the null hypothesis and say we have evidence that the two groups have different means—that they come from different parent distributions. When we reject H_0, we say that the difference is *statistically significant*. The threshold for accepting/rejecting H_0 is called α, usually with $\alpha = 0.05$ as a typical, if problematic, value. We'll discuss why 0.05 is problematic below.

The point to remember is that the *p*-value assumes the null hypothesis is true. It tells us the likelihood of a true H_0 giving us at least the difference we see, or greater, between the groups. If the *p*-value is small, that has two possible meanings: (1) the null hypothesis is false, or (2) a random sampling error has given us samples that fall outside what we might expect. Since the *p*-value assumes H_0 is true, a small *p*-value helps us believe less and less in (2) and boosts our confidence that (1) might be correct. However, the *p*-value alone cannot confirm (1); other knowledge needs to come into play.

I mentioned that using $\alpha = 0.05$ is problematic. The main reason it's problematic is that it's too generous; it leads to too many rejections of a true null hypothesis. According to James Berger and Thomas Sellke in their article "Testing a Point Null Hypothesis: The Irreconcilability of *P* Values and Evidence" (*Journal of the American Statistical Association*, 1987), when $\alpha = 0.05$, about 30 percent of true null hypotheses will be rejected. When we use something like $\alpha \leq 0.001$, the chance of falsely rejecting a true null hypothesis goes down to less than 3 percent. The moral of the story is that $p < 0.05$ is not magic and, frankly, is unconvincing for a single study. Look for highly significant *p*-values of at least 0.001 or, preferably, much smaller. At $p = 0.05$, all you have is a suggestion, and you should repeat the experiment. If repeated experiments all have a *p*-value around 0.05, then rejecting the null hypothesis begins to make sense.

Confidence Intervals

Along with a *p*-value, you'll often see *confidence intervals (CIs)*. The confidence interval gives bounds within which we believe the true population difference in the means will lie, with a given confidence for repeated samples of the two datasets we're comparing. Typically, we report 95 percent confidence intervals. Our hypothesis tests check for equality of means by asking if the difference of the sample means is zero or not. Therefore, any CI that includes zero signals to us that we cannot reject the null hypothesis.

For Welch's t-test, the degrees of freedom is

$$df = \frac{\left(\frac{s_1^2}{n_1} + \frac{s_2^2}{n_2}\right)^2}{\frac{(s_1^2/n_1)^2}{n_1-1} + \frac{(s_2^2/n_2)^2}{n_2-1}} \tag{4.7}$$

which we can use to calculate confidence intervals,

$$CI_\alpha = (\bar{x}_1 - \bar{x}_2) \pm t_{1-\alpha/2,df}\sqrt{\frac{s_1^2}{n_1} + \frac{s_2^2}{n_2}} \qquad (4.8)$$

where $t_{1-\alpha/2,df}$ is the *critical value*, and the t-value for the given confidence level (α) and the degrees of freedom, *df*, come from Equation 4.7.

How should we interpret the 95 percent confidence interval? There is a population value: the true difference between the group means. The 95 percent confidence interval is such that if we could draw repeated samples from the distribution that produced the two datasets, 95 percent of the calculated confidence intervals would contain the true difference between the means. It is *not* the range that includes the true difference in the means at 95 percent certainty.

Beyond checking if zero is in the CI, the CI is useful because its width tells us something about the magnitude of the effect. Here, the effect is related to the difference between the means. We may have a statistically significant difference based on the *p*-value, but the effect might be practically meaningless. The CI will be narrow when the effect is large because small CIs imply a narrow range encompassing the true effect. We'll see shortly how, when possible, to calculate another useful measure of effect.

Finally, a *p*-value less than α also will have a CI_α that does not include H_0. In other words, what the *p*-value tells us and what the confidence interval tells us track—they will not contradict each other.

Effect Size

It's one thing to have a statistically significant *p*-value. It's another for the difference represented by that *p*-value to be meaningful in the real world. A popular measure of the size of an effect, the *effect size*, is *Cohen's d*. For us, since we're using Welch's t-test, Cohen's *d* is found by calculating

$$d = \frac{\bar{x}_2 - \bar{x}_1}{\sqrt{\frac{1}{2}\left(s_1^2 + s_2^2\right)}} \qquad (4.9)$$

Cohen's *d* is usually interpreted subjectively, though we should report the numeric value as well. Subjectively, the size of the effect could be

d	Effect
0.2	Small
0.5	Medium
0.8	Large

Cohen's d makes sense. The difference between the means is a natural way to think about the effect. Scaling it by the mean variance puts it in a consistent range. From Equation 4.9, we see that a p-value corresponding to a statistically significant result might lead to a small effect that isn't of any true practical importance.

Evaluating the Test Scores

Let's put all of the above together to apply the t-test to our test data from Table 4-1. You'll find the code in the file *hypothesis.py*. We generate the datasets first:

```
np.random.seed(65535)
a = np.random.normal(85,6,50).astype("int32")
a[np.where(a > 100)] = 100
b = np.random.normal(82,7,50).astype("int32")
b[np.where(b > 100)] = 100
```

Once again, we're using a fixed NumPy pseudorandom number seed for repeatability. We make a a sample from a normal distribution with a mean of 85 and a standard deviation of 6.0. We select b from a normal distribution with a mean of 82 and a standard deviation of 7.0. For both, we cap any values over 100 to 100. These are test scores, after all, without extra credit.

We apply the t-test next:

```
from scipy.stats import ttest_ind
t,p = ttest_ind(a,b, equal_var=False)
print("(t=%0.5f, p=%0.5f)" % (t,p))
```

We get $(t = 2.40234, p = 0.01852)$. The t is the statistic, and p is the computed p-value. It's 0.019, which is less than 0.05 but only by a factor of two. We have a weak result telling us we might want to reject the null hypothesis and believe that the two groups, a and b, come from different distributions. Of course, we know they do because we generated them, but it's nice to see the test pointing in the right direction.

Notice that the function we import from SciPy is ttest_ind. This is the function to use for independent samples, which are not paired. Also, notice that we added equal_var=False to the call. This is how to use Welch's t-test, which doesn't assume that the variance between the two datasets is equal. We know they're not equal, since a uses a standard deviation of 6.0 while b uses 7.0.

To get the confidence intervals, we'll write a CI function, since NumPy and SciPy don't include one. The function directly implements Equations 4.7 and 4.8:

```
from scipy import stats
def CI(a, b, alpha=0.05):
    n1, n2 = len(a), len(b)
```

```
s1, s2 = np.std(a, ddof=1)**2, np.std(b, ddof=1)**2
df = (s1/n1 + s2/n2)**2 / ((s1/n1)**2/(n1-1) + (s2/n2)**2/(n2-1))
tc = stats.t.ppf(1 - alpha/2, df)
lo = (a.mean()-b.mean()) - tc*np.sqrt(s1/n1 + s2/n2)
hi = (a.mean()-b.mean()) + tc*np.sqrt(s1/n1 + s2/n2)
return lo, hi
```

The critical t value is given by calling stats.t.ppf, passing in the $\alpha/2$ value and the proper degrees of freedom, df. The critical t value is the 97.5 percent percentile value, for $\alpha = 0.05$, which is what the *percent point function (ppf)* returns. We divide by two to cover the tails of the t-distribution.

For our test example, the confidence interval is $[0.56105, 5.95895]$. Notice how this does not include zero, so the CI also indicates a statistically significant result. However, the range is rather large, so this is not a particularly robust result. The CI range can be difficult to interpret on its own, so, finally, let's calculate Cohen's d to see if it makes sense given the width of the confidence interval. In code, we implement Equation 4.9:

```
def Cohen_d(a,b):
    s1 = np.std(a, ddof=1)**2
    s2 = np.std(b, ddof=1)**2
    return (a.mean() - b.mean()) / np.sqrt(0.5*(s1+s2))
```

We get $d = 0.48047$, corresponding to a medium effect size.

The Mann-Whitney U Test

The t-test assumes the distribution of the source data is normal. If the data is not normally distributed, we should instead use a *nonparametric test*. Nonparametric tests make no assumptions about the underlying distribution of the data. The *Mann-Whitney U test*, sometimes called the *Wilcoxon rank-sum test*, is a nonparametric test to help decide if two different sets of data come from the same parent distribution. The Mann-Whitney U test does not rely directly on the values of the data, but instead uses the data's ranking.

The null hypothesis for this test is the following: the probability that a randomly selected value from Group 1 is larger than a randomly selected value from Group 2 is 0.5. Let's think a bit about that. If the data is from the same parent distribution, then we should expect any randomly selected pair of values from the two groups to show no preference as to which is larger than the other.

The alternative hypothesis is that the probability of a randomly selected value from Group 1 being larger than a randomly selected value from Group 2 is not 0.5. Notice, there is no statement as to the probability being greater or less than 0.5, only that it isn't 0.5; thus, the Mann-Whitney U test, as we'll use it, is two-sided.

The null hypothesis for the Mann-Whitney U test is not the same as the null hypothesis for the t-test. For the t-test, we're asking whether the means between the two groups are the same. (Really, we're asking if the difference in the means is zero.) However, if two sets of data *are* from different parent distributions, both null hypotheses are false, so we can use the Mann-Whitney U test in place of the t-test, especially when the underlying data is not normally distributed.

To generate U, the Mann-Whitney statistic, we first pool both sets of data and rank them. Ties are replaced with the mean between the tie value rank and the next rank value. We also keep track of the source group so we can separate the list of ranks again. The ranks, by group, are summed to give R_1 and R_2 (using the ranks from the pooled data). We calculate two values,

$$U_1 = n_1 n_2 + \frac{n_1(n_1 - 1)}{2} - R_1$$

$$U_2 = n_1 n_2 + \frac{n_2(n_2 - 1)}{2} - R_2$$

with the smaller called U, the test statistic. It's possible to generate a *p*-value from U, keeping in mind all the discussion above about the meaning and use of *p*-values. As before, n_1 and n_2 are the number of samples in the two groups. The Mann-Whitney U test requires the smaller of these two numbers to be at least 21 samples. If you don't have that many, the results may not be reliable when using the SciPy `mannwhitneyu` function.

We can run the Mann-Whitney U test on our test data from Table 4-1,

```
from scipy.stats import mannwhitneyu
u,p = mannwhitneyu(a,b)
print("(U=%0.5f, p=%0.5f)" % (u,p))
```

with a and b as we used above for the t-test. This gives us ($U = 997.00000$, $p = 0.04058$). The *p*-value is barely below the minimum threshold of 0.05.

The means of a and b are 85 and 82, respectively. What happens to the *p*-values if we make the mean value of b 83 or 81? Changing the mean of b means changing the first argument to `np.random.normal`. Doing this gives us Table 4-2, where I've included all results for completeness.

Table 4-2: Mann-Whitney U Test and t-test Results for the Simulated Test Scores with Different Means ($n_1 = n_2 = 50$)

Means	Mann-Whitney U	t-test
85 vs. 83	(U=1104.50000, p=0.15839)	(t=1.66543, p=0.09959)
85 vs. 82	(U=997.00000, p=0.04058)	(t=2.40234, p=0.01852)
85 vs. 81	(U=883.50000, p=0.00575)	(t=3.13925, p=0.00234)

Table 4-2 should make sense to us. When the means are close, it's harder to tell them apart, so we expect larger *p*-values. Recall how we have only 50 samples in each group. As the difference between the means increases, the *p*-values go down. A difference of three in the means leads to barely significant *p*-values. When the difference is larger still, the *p*-values become truly significant—again, as we expect.

The analysis above begs the question: for a small difference in the means between the two groups, how do the *p*-values change as a function of the sample size?

Figure 4-8 shows the *p*-value (mean ± standard error) over 25 runs for both the Mann-Whitney U test and the t-test as a function of sample size for the case where the means are 85 and 84.

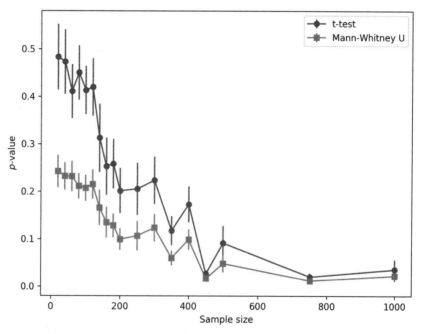

Figure 4-8: Mean p-value as a function of sample size for a difference in the sample means of one, $\bar{x}_1 = 85$, $\bar{x}_2 = 84$

Small datasets make it difficult to differentiate between cases when the difference in the means is small. We also see that larger sample sizes reveal the difference, regardless of the test. It is interesting that in Figure 4-8, the Mann-Whitney U *p*-value is less than that of the t-test even though the underlying data is normally distributed. Conventional wisdom states that it's usually the other way around.

Figure 4-8 is an object lesson in the power of large-sample tests to detect real differences. When the sample size is large enough, a weak difference becomes significant. However, we need to balance this with the effect size. When we have 1,000 samples in each group, we have a statistically significant *p*-value, but we also have a Cohen's *d* of about 0.13, signaling a weak effect. A large sample study might find a significant effect that is so weak as to be practically meaningless.

Summary

This chapter touched on the key aspects of statistics you'll encounter during your sojourn through the world of deep learning. Specifically, we learned about different types of data and how to ensure the data is useful for building models. We then learned about summary statistics and saw examples that used them to help us understand a dataset. Understanding our data is key to successful deep learning. We investigated the different types of means, learned about measures of variation, and saw the utility of visualizing the data via box plots.

Missing data is a bane of deep learning. In this chapter, we investigated how to compensate for missing data. Next, we discussed correlation, how to detect and measure the relationships between elements of a dataset. Finally, we introduced hypothesis testing. Restricting ourselves to the most likely scenario we'll encounter in deep learning, we learned how to apply both the t-test and the Mann-Whitney U test. Hypothesis testing introduced us to the p-value. We saw examples of it and discussed how to interpret it correctly.

In the next chapter we'll leave statistics behind and dive headfirst into the world of linear algebra. Linear algebra is how we implement neural networks.

5

LINEAR ALGEBRA

Formally, linear algebra is the study of linear equations, in which the highest power of the variable is one. However, for our purposes, *linear algebra* refers to multidimensional mathematical objects—like vectors and matrices—and operations on them. This is how linear algebra is typically applied in deep learning, and how data is manipulated in programs that implement deep learning algorithms. By making this distinction, we are throwing away a massive amount of fascinating mathematics, but as our goal is to understand the mathematics used and applied in deep learning, we can hopefully be forgiven.

In this chapter, I'll introduce the objects used in deep learning, specifically scalars, vectors, matrices, and tensors. As we'll see, all of these objects are actually tensors of various orders. We'll discuss tensors from a mathematical, notational perspective and then experiment with them using NumPy. NumPy was explicitly designed to add multidimensional arrays to Python, and they are good, though incomplete, analogues for the mathematical objects we'll work with in this chapter.

We'll spend the bulk of the chapter learning how to do arithmetic with tensors, which is of fundamental importance in deep learning. Most of the effort in implementing highly performant deep learning toolkits involves finding ways to do arithmetic with tensors as efficiently as possible.

Scalars, Vectors, Matrices, and Tensors

Let's introduce our cast of characters. I'll relate them to Python variables and NumPy arrays to show how we'll implement these objects in code. Then I'll present a handy conceptual mapping between tensors and geometry.

Scalars

Even if you're not familiar with the word, you've known what a scalar is since the day you first learned to count. A *scalar* is just a number, like 7, 42, or π. In expressions, we'll use x to mean a scalar, that is, the ordinary notation used for variables. To a computer, a scalar is a simple numeric variable:

```
>>> s = 66
>>> s
66
```

Vectors

A *vector* is a 1D array of numbers. Mathematically, a vector has an orientation, either horizontal or vertical. If horizontal, it's a *row vector*. For example,

$$x = [x_0 \ x_1 \ x_2] \tag{5.1}$$

is a row vector of three elements or components. Note, we'll use x, a lowercase letter in bold, to mean a vector.

Mathematically, vectors are usually assumed to be *column vectors*,

$$y = \begin{pmatrix} y_0 \\ y_1 \\ y_2 \\ y_3 \end{pmatrix} \tag{5.2}$$

where y has four components, making it a four-dimensional (4D) vector. Notice that in Equation 5.1 we used square brackets, whereas in Equation 5.2 we used parentheses. Either notation is acceptable.

In code, we usually implement vectors as 1D arrays:

```
>>> import numpy as np
>>> x = np.array([1,2,3])
>>> print(x)
[1 2 3]
>>> print(x.reshape((3,1)))
[[1]
 [2]
 [3]]
```

Here, we've used reshape to turn the three-element row vector into a column vector of three rows and one column.

The components of a vector are often interpreted as lengths along a set of coordinate axes. For example, a three-component vector might be used to represent a point in 3D space. In this vector,

$$x = [x, y, z]$$

x could be the length along the x-axis, y the length along the y-axis, and z the length along the z-axis. These are the Cartesian coordinates and serve to uniquely identify all points in 3D space.

However, in deep learning, and machine learning in general, the components of a vector are often unrelated to each other in any strict geometric sense. Rather, they're used to represent *features*, qualities of some sample that the model will use to attempt to arrive at a useful output, like a class label, or a regression value. That said, the vector representing the collection of features, called the *feature vector*, is sometimes thought about geometrically. For example, some machine learning models, like k-nearest neighbors, interpret the vector as representing some coordinate in geometric space.

You'll often hear deep learning people discuss the *feature space* of a problem. The feature space refers to the set of possible inputs. The training set for a model needs to accurately represent the feature space of the possible inputs the model will encounter when used. In this sense, the feature vector is a point, a location in this n-dimensional space where n is the number of features in the feature vector.

Matrices

A *matrix* is a 2D array of numbers:

$$A = \begin{bmatrix} a_{00} & a_{01} & a_{02} & a_{03} \\ a_{10} & a_{11} & a_{12} & a_{13} \\ a_{20} & a_{21} & a_{22} & a_{23} \end{bmatrix}$$

The elements of A are subscripted by the row number and column number. The matrix A has three rows and four columns, so we say that it's a 3×4 matrix, where 3×4 is the *order* of the matrix. Notice that A uses subscripts starting with 0. Math texts often begin with 1, but increasingly, they're using 0 so that there isn't an offset between the math notation and the computer representation of the matrix. Note, also, that we'll use A, an uppercase letter in bold, to mean a matrix.

In code, matrices are represented as 2D arrays:

```
>>> A = np.array([[1,2,3],[4,5,6],[7,8,9]])
>>> print(A)
[[1 2 3]
 [4 5 6]
 [7 8 9]]
>>> print(np.arange(12).reshape((3,4)))
[[ 0  1  2  3]
 [ 4  5  6  7]
 [ 8  9 10 11]]
```

To get element a_{12} of A in Python, we write A[1,2]. Notice that when we printed the arrays, there was an extra [and] around them. NumPy uses these brackets to indicate that the 2D array can be thought of as a row vector in which each element is itself a vector. In Python-speak, this means that a matrix can be thought of as a list of sublists in which each sublist is of the same length. Of course, this is exactly how we defined A to begin with.

We can think of vectors as matrices with a single row or column. A column vector with three elements is a 3×1 matrix: it has three rows and one column. Similarly, a row vector of four elements acts like a 1×4 matrix: it has one row and four columns. We'll make use of this observation later.

Tensors

A scalar has no dimensions, a vector has one, and a matrix has two. As you might suspect, we don't need to stop there. A mathematical object with more than two dimensions is colloquially referred to as a *tensor*. When necessary, we'll represent tensors like this: T, as a sans serif capital letter.

The number of dimensions a tensor has defines its *order*, which is not to be confused with the order of a matrix. A 3D tensor has order 3. A matrix is a tensor of order 2. A vector is an order-1 tensor, and a scalar is an order-0 tensor. When we discuss the flow of data through a deep neural network in Chapter 9, we'll see that many toolkits use tensors of order 4 (or more).

In Python, NumPy arrays with three or more dimensions are used to implement tensors. For example, we can define an order-3 tensor in Python as shown below:

```
>>> t = np.arange(36).reshape((3,3,4))
>>> print(t)
[[[ 0  1  2  3]
  [ 4  5  6  7]
  [ 8  9 10 11]]

 [[12 13 14 15]
  [16 17 18 19]
  [20 21 22 23]]

 [[24 25 26 27]
  [28 29 30 31]
  [32 33 34 35]]]
```

Here, we use np.arange to define t to be a vector of 36 elements holding the numbers $0 \ldots 35$. Then, we immediately reshape the vector into a tensor of $3 \times 3 \times 4$ elements ($3 \times 3 \times 4 = 36$). One way to think of a $3 \times 3 \times 4$ tensor is that it contains a stack of three 3×4 images. If we keep this in mind, the following statements make sense:

```
>>> print(t[0])
[[ 0  1  2  3]
 [ 4  5  6  7]
 [ 8  9 10 11]]
>>> print(t[0,1])
[4 5 6 7]
>>> print(t[0,1,2])
6
```

Asking for t[0] will return the first 3×4 *image* in the stack. Asking for t[0,1], then, should return the second row of the first image, which it does. Finally, we get to an individual element of t by asking for the image number (0), the row number (1), and the element of that row (2).

Assigning the dimensions of a tensor to successively smaller collections of something is a handy way to keep the meaning of the dimensions in mind. For example, we can define an order-5 tensor like so:

```
>>> w = np.zeros((9,9,9,9,9))
>>> w[4,1,2,0,1]
0.0
```

But, what does asking for w[4,1,2,0,1] mean? The exact meaning depends on the application. For example, we might think of w as representing a bookcase. The first index selects the shelf, and the second selects the book on the shelf. Then, the third index selects the page within the book, and the fourth selects the line on the page. The final index selects the word on the line. Therefore, w[4,1,2,0,1] is asking for the second word of the first line of the third page of the second book on the fifth shelf of the bookcase, understood by reading the indices from right to left.

The bookcase analogy does have its limitations. NumPy arrays have fixed dimensions, meaning that if w is a bookcase, there are nine shelves, and each shelf has *exactly* nine books. Likewise, each book has exactly nine pages, and each page has nine lines. Finally, each line has precisely nine words. NumPy arrays ordinarily use contiguous memory in the computer, so the size of each dimension is fixed when the array is defined. Doing so, and selecting the specific data type, like unsigned integer, makes locating an element of the array an indexing operation using a simple formula to compute an offset from a base memory address. This is what makes NumPy arrays so much faster than Python lists.

Any tensor of less than order n can be represented as an order-n tensor by supplying the missing dimensions of length one. We saw an example of this above when I said that an m-component vector could be thought of as a $1 \times m$ or an $m \times 1$ matrix. The order-1 tensor (the vector) is turned into an order-2 tensor (matrix) by adding a missing dimension of length one.

As an extreme example, we can treat a scalar (order-0 tensor) as an order-5 tensor, like this:

```
>>> t = np.array(42).reshape((1,1,1,1,1))
>>> print(t)
[[[[[42]]]]]
>>> t.shape
(1, 1, 1, 1, 1)
>>> t[0,0,0,0,0]
42
```

Here, we reshape the scalar 42 into an order-5 tensor (a five-dimensional [5D] array) with length one on each axis. Notice that NumPy tells us that the tensor t has five dimensions with the [[[[[and]]]]] around 42. Asking for the shape of t confirms that it is a 5D tensor. Finally, as a tensor, we can get the value of the single element it contains by specifying all the dimensions with t[0,0,0,0,0]. We'll often use this trick of adding new dimensions of length one. In fact, in NumPy, there is a way to do this directly, which you'll see when using deep learning toolkits:

```
>>> t = np.array([[1,2,3],[4,5,6]])
>>> print(t)
[[1 2 3]
 [4 5 6]]
>>> w = t[np.newaxis,:,:]
```

```
>>> w.shape
(1, 2, 3)
>>> print(w)
[[[1 2 3]
  [4 5 6]]]
```

Here, we've turned t, an order-2 tensor (a matrix), into an order-3 tensor by using np.newaxis to create a new axis of length one. That's why w.shape returns (1,2,3) and not (2,3), as it would for t.

There are analogues between tensors up to order-3 and geometry that are helpful in visualizing the relationships between the different orders:

Order (dimensions)	Tensor name	Geometric name
0	Scalar	Point
1	Vector	Line
2	Matrix	Plane
3	Tensor	Volume

Notice, I used *tensor* in its common sense in the table. There seems to be no standardized name for an order-3 tensor.

In this section, we defined the mathematical objects of deep learning in relation to multidimensional arrays, since that's how they are implemented in code. We've thrown away a lot of mathematics by doing this, but we've preserved what we need to understand deep learning. Let's move on now and see how to use tensors in expressions.

Arithmetic with Tensors

The purpose of this section is to detail operations on tensors, with special emphasis on tensors of order-1 (vectors) and order-2 (matrices). We'll assume operations with scalars are well in hand at this point.

We'll start with what I'm calling *array operations*, by which I mean the element-wise operations that toolkits like NumPy perform on arrays of all dimensions. Then we'll move on to operations particular to vectors. This sets the stage for the critical topic of matrix multiplication. Finally, we'll discuss block matrices.

Array Operations

The way we've used the NumPy toolkit so far has shown us that all the normal scalar arithmetic operations translate directly into the world of multidimensional arrays. This includes standard operations like addition, subtraction, multiplication, division, and exponentiation, as well as the application of functions to an array. In all of these cases, the scalar operation is applied element-wise to each element of the array. The examples here will set the tone for the rest of this section and will also let us explore some NumPy broadcasting rules that we haven't called out yet.

Let's first define some arrays to work with:

```
>>> a = np.array([[1,2,3],[4,5,6]])
>>> b = np.array([[7,8,9],[10,11,12]])
>>> c = np.array([10,100,1000])
>>> d = np.array([10,11])
>>> print(a)
[[1 2 3]
 [4 5 6]]
>>> print(b)
[[ 7  8  9]
 [10 11 12]]
>>> print(c)
[  10  100 1000]
>>> print(d)
[10 11]
```

Element-wise arithmetic is straightforward for arrays with dimensions that match:

```
>>> print(a+b)
[[ 8 10 12]
 [14 16 18]]
>>> print(a-b)
[[-6 -6 -6]
 [-6 -6 -6]]
>>> print(a*b)
[[ 7 16 27]
 [40 55 72]]
>>> print(a/b)
[[0.14285714 0.25       0.33333333]
 [0.4        0.45454545 0.5       ]]
>>> print(b**a)
[[      7      64     729]
 [  10000  161051 2985984]]
```

These results are all easy enough to interpret; NumPy applies the desired operation to the corresponding elements of each array. Element-wise multiplication on two matrices (a and b) is often known as the *Hadamard product*. (You'll encounter this term from time to time in the deep learning literature.)

The NumPy toolkit extends the idea of element-wise operations into what it calls *broadcasting*. When broadcasting, NumPy applies rules, which we'll see via examples, where one array is passed over another to produce a meaningful output.

We've already encountered a form of broadcasting when operating on an array with a scalar. In that case, the scalar value was broadcast to every value of the array.

For our first example, even though a is a 2 × 3 matrix, NumPy allows operations on it with c, a three-component vector, by applying broadcasting:

```
>>> print(a+c)
[[   11  102 1003]
 [   14  105 1006]]
>>> print(c*a)
[[   10  200 3000]
 [   40  500 6000]]
>>> print(a/c)
[[0.1   0.02  0.003]
 [0.4   0.05  0.006]]
```

Here, the three-component vector, c, has been broadcast over the rows of the 2 × 3 matrix, a. NumPy recognized that the last dimensions of a and c were both three, so the vector could be passed over the matrix to produce the given output. When looking at deep learning code, much of which is in Python, you'll see situations like this. At times, some thought is necessary, along with some experimentation at the Python prompt, to understand what's happening.

Can we broadcast d, a two-component vector, over a, a 2 × 3 matrix? If we try to do so the same way we broadcast c over a, we'll fail:

```
>>> print(a+d)
Traceback (most recent call last):
  File "<stdin>", line 1, in <module>
ValueError: operands could not be broadcast together with shapes (2,3) (2,)
```

However, the broadcasting rules for NumPy accommodate dimensions of length one. The shape of d is 2; it's a two-element vector. If we reshape d so that it's a 2D array with shape 2 × 1, we'll give NumPy what it needs:

```
>>> d = d.reshape((2,1))
>>> d.shape
(2, 1)
>>> print(a+d)
[[11 12 13]
 [15 16 17]]
```

We now see that Numpy has added d across the columns of a.

Let's return to the world of mathematics and look at operations on vectors.

Vector Operations

Vectors are represented in code as a collection of numbers that can be interpreted as values along a set of coordinate axes. Here, we'll define several operations that are unique to vectors.

Magnitude

Geometrically, we can understand vectors as having a direction and a length. They're often drawn as arrows, and we'll see an example of a vector plot in Chapter 6. People speak of the length of a vector as its *magnitude*. Therefore, the first vector operation we'll consider is calculating its magnitude. For a vector, x, with n components, the formula for its magnitude is

$$\|x\| = \sqrt{x_0^2 + x_1^2 + \cdots + x_{n-1}^2} \qquad (5.3)$$

In Equation 5.3, the double vertical bars around the vector represent its magnitude. You'll often see people use single bars here as well. Single bars are also used for absolute value; we usually rely on context to tell the difference between the two.

Where did Equation 5.3 come from? Consider a vector in 2D, $x = (x, y)$. If x and y are lengths along the x-axis and y-axis, respectively, we see that x and y form the sides of a right triangle. The length of the hypotenuse of this right triangle is the length of the vector. Therefore, according to Pythagoras, and the Babylonians long before him, this length is $\sqrt{x^2 + y^2}$, which, generalized to n dimensions, becomes Equation 5.3.

Unit Vectors

Now that we can calculate the magnitude of a vector, we can introduce a useful form of a vector known as a *unit vector*. If we divide the components of a vector by its magnitude, we're left with a vector that points in the same direction as the original vector but has a magnitude of one. This is the unit vector. For a vector, v, the unit vector in the same direction is

$$\hat{v} = \frac{v}{\|v\|}$$

where the hat over the vector serves to identify it as a unit vector. Let's see a concrete example. Our example vector is $v = (2, -4, 3)$. Therefore, the unit vector in the same direction as v is

$$\hat{v} = \frac{(2, -4, 3)}{\sqrt{2^2 + (-4)^2 + 3^2}} = \left(\frac{2}{\sqrt{29}}, \frac{-4}{\sqrt{29}}, \frac{3}{\sqrt{29}} \right) \approx (0.3714, -0.7428, 0.5571)$$

In code, we calculate the unit vector as the following:

```
>>> v = np.array((2, -4, 3))
>>> u = v / np.sqrt((v*v).sum())
>>> print(u)
[ 0.37139068 -0.74278135  0.55708601 ]
```

Here, we make use of the fact that to square each element of v, we multiply it by itself, element-wise, and then add the components together by calling sum to get the magnitude squared.

Vector Transpose

We mentioned earlier that row vectors can be thought of as $1 \times n$ matrices, while column vectors are $n \times 1$ matrices. The act of changing a row vector into a column vector and vice versa is known as taking the *transpose*. We'll see in Chapter 6 that the transpose also applies to matrices. Notationally, we denote the vector transpose of y as $y^\top$. Therefore, we have

$$x = [x_0 \ x_1 \ x_2]$$

$$x^\top = \begin{bmatrix} x_0 \\ x_1 \\ x_2 \end{bmatrix}$$

$$y = \begin{bmatrix} y_0 \\ y_1 \\ y_2 \end{bmatrix}$$

$$y^\top = [y_0 \ y_1 \ y_2]$$

$$(z^\top)^\top = z$$

Of course, we're not limited to just three components.

In code, we transpose vectors in several ways. As we saw above, we can use reshape to reshape the vector into a $1 \times n$ or $n \times 1$ matrix. We can also call the transpose method on the vector, with some care, or use the transpose shorthand. Let's see examples of all of these approaches. First, let's define a NumPy vector and see how reshape turns it into a 3×1 column vector and a 1×3 row vector, as opposed to a plain vector of three elements:

```
>>> v = np.array([1,2,3])
>>> print(v)
[1 2 3]
>>> print(v.reshape((3,1)))
[[1]
 [2]
 [3]]
>>> print(v.reshape((1,3)))
[[1 2 3]]
```

Notice the difference between the first print(v) and the last after calling reshape((1,3)). The output now has an extra set of brackets around it to indicate the leading dimension of one.

Next, we apply the transpose operation on v:

```
>>> print(v.transpose())
[1 2 3]
>>> print(v.T)
[1 2 3]
```

Here, we see that calling transpose or T changes nothing about v. This is because the shape of v is simply 3, not (1,3) or (3,1). If we explicitly alter v to be a 1 × 3 matrix, we see that transpose and T have the desired effect:

```
>>> v = v.reshape((1,3))
>>> print(v.transpose())
[[1]
 [2]
 [3]]
>>> print(v.T)
[[1]
 [2]
 [3]]
```

Here, v goes from being a row vector to a column vector, as we expect. The lesson, then, is to be careful about the actual dimensionality of vectors in NumPy code. Most of the time, we can be sloppy, but sometimes we need to be explicit and care about the distinction between plain vectors, row vectors, and column vectors.

Inner Product

Perhaps the most common vector operation is the *inner product*, or, as it is frequently called, the *dot product*. Notationally, the inner product between two vectors is written as

$$a \cdot b = \langle a, b \rangle = a^\top b$$

$$= \sum_{k=0}^{n-1} a_k b_k \tag{5.4}$$

$$= \|a\|\|b\| \cos \theta \tag{5.5}$$

Here, θ is the angle between the two vectors if they're interpreted geometrically. The result of the inner product is a scalar. The $\langle a, b \rangle$ notation is seen frequently, though the $a \cdot b$ dot notation seems more common in the deep learning literature. The $a^\top b$ matrix multiplication notation explicitly calls out how to calculate the inner product, but we'll wait until we discuss matrix multiplication to explain its meaning. For the present, the summation tells us what we need to know: the inner product of two vectors of length n is the sum of the products of the n components.

The inner product of a vector with itself is the magnitude squared:

$$a \cdot a = \|a\|^2$$

The inner product is commutative,

$$a \cdot b = b \cdot a$$

and distributive,

$$a \cdot (b + c) = a \cdot b + a \cdot c$$

but not associative, as the output of the first inner product is a scalar, not a vector, and multiplying a vector by a scalar is not an inner product.

Finally, notice that the inner product is zero when the angle between the vectors is 90 degrees; this is because $\cos \theta$ is zero (Equation 5.5). This means the two vectors are perpendicular, or *orthogonal*, to each other.

Let's look at some examples of the inner product. First, we'll be literal and implement Equation 5.4 explicitly:

```
>>> a = np.array([1,2,3,4])
>>> b = np.array([5,6,7,8])
>>> def inner(a,b):
...     s = 0.0
...     for i in range(len(a)):
...         s += a[i]*b[i]
...     return s
...
>>> inner(a,b)
70.0
```

However, since a and b are NumPy arrays, we know we can be more efficient:

```
>>> (a*b).sum()
70
```

Or, probably most efficient of all, we'll let NumPy do it for us by using np.dot:

```
>>> np.dot(a,b)
70
```

You'll see np.dot frequently in deep learning code. It can do more than calculate the inner product, as we'll see below.

Equation 5.5 tells us that the angle between two vectors is

$$\theta = \cos^{-1} \frac{a \cdot b}{\|a\|\|b\|}$$

In the code, this could be calculated as

```
>>> A = np.sqrt(np.dot(a,a))
>>> B = np.sqrt(np.dot(b,b))
>>> t = np.arccos(np.dot(a,b)/(A*B))
>>> t*(180/np.pi)
14.335170291600924
```

This tells us that the angle between a and b is approximately 14° after converting t from radians.

If we consider vectors in 3D space, we see that the dot product between orthogonal vectors is zero, implying that the angle between them is 90°:

```
>>> a = np.array([1,0,0])
>>> b = np.array([0,1,0])
>>> np.dot(a,b)
0
>>> t = np.arccos(0)
>>> t*(180/np.pi)
90.0
```

This is true because a is a unit vector along the x-axis, b is a unit vector along the y-axis, and we know there's a right angle between them.

With the inner product in our toolkit, let's see how we can use it to project one vector onto another.

Projection

The projection of one vector onto another calculates the amount of the first vector that's in the direction of the second. The projection of a onto b is

$$\mathbf{proj}_b a = \frac{a \cdot b}{\|b\|^2} b$$

Figure 5-1 shows graphically what projection means for 2D vectors.

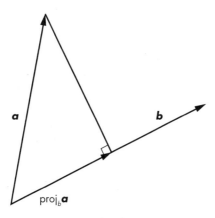

Figure 5-1: A graphical representation
of the projection of a onto b in 2D

Projection finds the component of *a* in the direction of *b*. Note the projection of *a* onto *b* is not the same as the projection of *b* onto *a*.

Because we use an inner product in the numerator, we can see that the projection of a vector onto another vector that's orthogonal to it is zero. No component of the first vector is in the direction of the second. Think again of the x-axis and y-axis. The entire reason we use Cartesian coordinates is because the two axes, or three in 3D space, are all mutually orthogonal; no part of one is in the direction of the others. This lets us specify any point, and the vector from the origin to that point, by specifying the components along these axes. We'll see this breaking up of an object into mutually orthogonal components later when we discuss eigenvectors and PCA in Chapter 6.

In code, calculating the projection is straightforward:

```
>>> a = np.array([1,1])
>>> b = np.array([1,0])
>>> p = (np.dot(a,b)/np.dot(b,b))*b
>>> print(p)
[1. 0.]
>>> c = np.array([-1,1])
>>> p = (np.dot(c,b)/np.dot(b,b))*b
>>> print(p)
[-1. -0.]
```

In the first example, *a* points in the direction 45° up from the x-axis, while *b* points along the x-axis. We'd then expect the projection of *a* to be along the x-axis, which it is (p). In the second example, *c* points in the direction 135° = 90° + 45° from the x-axis. Therefore, we'd expect the component of *c* along *b* to be along the x-axis but in the opposite direction from *b*, which it is.

NOTE *Projecting c along b returned a y-axis component of −0. The negative sign is a quirk of the IEEE 754 representation used for floating-point numbers. The significand (mantissa) of the internal representation is zero, but the sign can still be specified, leading to an output of negative zero from time to time. For a detailed explanation of computer number formats, including floating-point, please see my book,* Numbers and Computers *(Springer-Verlag, 2017).*

Let's move on now to consider the outer product of two vectors.

Outer Product

The inner product of two vectors returned a scalar value. The *outer product* of two vectors instead returns a matrix. Note that unlike the inner product, the outer product does not require the two vectors to have the same number of components. Specifically, for vectors *a* of *m* components and *b* of *n* components, the outer product is the matrix formed by multiplying each element of *a* by each element of *b*, as shown next.

$$\boldsymbol{a} \otimes \boldsymbol{b} = \boldsymbol{ab}^\top = \begin{bmatrix} a_0 b_0 & a_0 b_1 & a_0 b_2 & \cdots & a_0 b_{n-1} \\ a_1 b_0 & a_1 b_1 & a_1 b_2 & \cdots & a_1 b_{n-1} \\ \vdots & \vdots & \vdots & \cdots & \vdots \\ a_{m-1} b_0 & a_{m-1} b_1 & a_{m-1} b_2 & \cdots & a_{m-1} b_{n-1} \end{bmatrix}$$

The $\boldsymbol{ab}^\top$ notation is how to calculate the outer product via matrix multiplication. Notice that this notation is not the same as the inner product, $\boldsymbol{a}^\top \boldsymbol{b}$, and that it assumes $\boldsymbol{a}$ and $\boldsymbol{b}$ to be column vectors. No operator symbol is consistently used for the outer product, primarily because it's so easily specified via matrix multiplication and because it's less common than the dot product. However, $\otimes$ seems the most commonly used when the outer product is presented with a binary operator.

In code, NumPy has kindly provided an outer product function for us:

```
>>> a = np.array([1,2,3,4])
>>> b = np.array([5,6,7,8])
>>> np.dot(a,b)
70
>>> np.outer(a,b)
array([[ 5,  6,  7,  8],
       [10, 12, 14, 16],
       [15, 18, 21, 24],
       [20, 24, 28, 32]])
```

We used a and b above when discussing the inner product. As expected, np.dot gives us a scalar output for $\boldsymbol{a} \cdot \boldsymbol{b}$. However, the np.outer function returns a 4×4 matrix, where we see that each row is vector b multiplied successively by each element of vector a, first 1, then 2, then 3, and finally 4. Therefore, each element of a has multiplied each element of b. The resulting matrix is 4×4 because both a and b have four elements.

THE CARTESIAN PRODUCT

There is a direct analogue between the outer product of two vectors and the Cartesian product of two sets, A and B. The *Cartesian product* is a new set, each element of which is one of the possible pairings of elements from A and B. So, if A={1,2,3,4} and B={5,6,7,8}, the Cartesian product can be written as

$$A \times B = \{(a,b) \mid a \in A \text{ and } b \in B\}$$
$$= \begin{pmatrix} (1,5) & (1,6) & (1,7) & (1,8) \\ (2,5) & (2,6) & (2,7) & (2,8) \\ (3,5) & (3,6) & (3,7) & (3,8) \\ (4,5) & (4,6) & (4,7) & (4,8) \end{pmatrix}$$

Here, we see that if we replace each entry with the product of the pair, we get the corresponding vector product we saw above with NumPy np.outer. Also, note that × is typically used for the Cartesian product when working with sets.

The ability of the outer product to mix all combinations of its inputs has been used in deep learning for neural collaborative filtering and visual question answering applications. These functions are performed by advanced networks that make recommendations or answer text questions about an image. The outer product appears as a mixing of two different embedding vectors. *Embeddings* are the vectors generated by lower layers of a network, for example, the next to last fully connected layer before the softmax layer's output of a traditional convolutional neural network (CNN). The embedding layer is usually viewed as having learned a new representation of the network input. It can be thought of as mapping complex inputs, like images, to a reduced space of several hundred to several thousands of dimensions.

Cross Product

Our final vector-vector operator is the *cross product*. This operator is only defined for 3D space ($\mathbb{R}^3$). The cross product of a and b is a new vector that is perpendicular to the plane containing a and b. Note, this does not imply that a and b are themselves perpendicular. The cross product is defined as

$$a \times b = \|a\|\|b\| \sin(\theta)\hat{n}$$

$$= (a_1 b_2 - a_2 b_1, a_0 b_2 - a_2 b_0, a_0 b_1 - a_1 b_0) \tag{5.6}$$

where $\hat{n}$ is a unit vector and θ is the angle between a and b. The direction of $\hat{n}$ is given by the *right-hand rule*. With your right hand, point your index finger in the direction of a and your middle finger in the direction of b. Then, your thumb will be pointing in the direction of $\hat{n}$. Equation 5.6 gives the actual $\mathbb{R}^3$ components of the cross product vector.

NumPy implements the cross product via np.cross:

```
>>> a = np.array([1,0,0])
>>> b = np.array([0,1,0])
>>> print(np.cross(a,b))
[0 0 1]
>>> c = np.array([1,1,0])
>>> print(np.cross(a,c))
[0 0 1]
```

In the first example, a points along the x-axis and b along the y-axis. Therefore, we expect the cross product to be perpendicular to these axes, and it is: the cross product points along the z-axis. The second example shows that it doesn't matter if a and b are perpendicular to each other. Here, c is at a 45° angle to the x-axis, but a and c are still in the xy-plane. Therefore, the cross product is still along the z-axis.

The definition of the cross product involves $\sin \theta$, while the inner product uses $\cos \theta$. The inner product is zero when the two vectors are orthogonal to each other. The cross product, on the other hand, is zero when the two vectors are in the same direction and is maximized when the vectors are perpendicular. The second NumPy example above works out because the magnitude of c is $\sqrt{2}$ and $\sin 45° = \sqrt{2}/2 = 1/\sqrt{2}$. As a result, the $\sqrt{2}$ factors cancel out to leave a magnitude of 1 for the cross product because a is a unit vector.

The cross product is widely used in physics and other sciences but is less often used in deep learning because of its restriction to 3D space. Nonetheless, you should be familiar with it if you're going to tackle the deep learning literature.

This concludes our look at vector-vector operations. Let's leave the 1D world and move on to consider the most important operation for all deep learning: matrix multiplication.

Matrix Multiplication

In the previous section, we saw how to multiply two vectors in various ways: Hadamard product, inner (dot) product, outer product, and cross product. In this section, we'll investigate multiplication of matrices, recalling that row and column vectors are themselves matrices with one row or column.

Properties of Matrix Multiplication

We'll define the matrix product operation shortly, but, before we do, let's look at the properties of matrix multiplication. Let A, B, and C be matrices. Then, following the algebra convention of multiplying symbols by placing them next to each other,

$$(AB)C = A(BC)$$

meaning matrix multiplication is associative. Second, matrix multiplication is distributive:

$$A(B + C) = AB + AC \tag{5.7}$$

$$(A + B)C = AC + BC \tag{5.8}$$

However, in general, matrix multiplication is *not* commutative:

$$AB \neq BA$$

As you can see in Equation 5.8, matrix multiplication over addition from the right produces a different result than matrix multiplication over addition from the left, as shown in Equation 5.7. This explains why we showed both Equation 5.7 and Equation 5.8; matrix multiplication can be performed from the left or the right, and the result will be different.

How to Multiply Two Matrices

To calculate AB, knowing that A must be on the left of B, we first need to verify that the matrices are compatible. It's only possible to multiply two matrices if the number of columns in A is the same as the number of rows in B. Therefore, if A is an $n \times m$ matrix and B is an $m \times k$ matrix, then the product, AB, can be found and will be a new $n \times k$ matrix.

To calculate the product, we perform a series of inner product multiplications between the row vectors of A and the column vectors of B. Figure 5-2 illustrates the process for a 3×3 matrix A and a 3×2 matrix B.

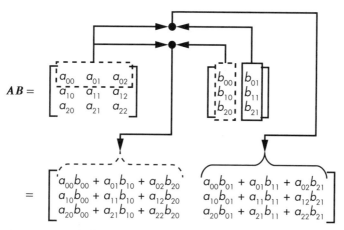

Figure 5-2: Multiplying a 3×3 matrix by a 3×2 matrix

In Figure 5-2, the first row of the output matrix is found by computing the inner product of the first row of A with each of the columns of B. The first element of the output matrix is shown where the first row of A is multiplied by the first column of B. The remaining first row of the output matrix is found by repeating the dot product of the first row of A by the remaining column of B.

Let's present a worked example with actual numbers for the matrices in Figure 5-2:

$$AB = \begin{bmatrix} 1 & 2 & 3 \\ 4 & 5 & 6 \\ 7 & 8 & 9 \end{bmatrix} \begin{bmatrix} 11 & 22 \\ 33 & 44 \\ 55 & 66 \end{bmatrix}$$

$$= \begin{bmatrix} (1)11 + (2)33 + (3)55 & (1)22 + (2)44 + (3)66 \\ (4)11 + (5)33 + (6)55 & (4)22 + (5)44 + (6)66 \\ (7)11 + (8)33 + (9)55 & (7)22 + (8)44 + (9)66 \end{bmatrix}$$

$$= \begin{bmatrix} 242 & 308 \\ 539 & 704 \\ 836 & 1100 \end{bmatrix}$$

Notice, AB is defined, but BA is not, because we can't multiply a 3×2 matrix by a 3×3 matrix. The number of columns in B needs to be the same as the number of rows in A.

Another way to think of matrix multiplication is by considering what goes into making up each of the output matrix elements. For example, if A is $n \times m$ and B is $m \times p$, we know that the matrix product exists as an $n \times p$ matrix, C. We find the output elements by computing

$$c_{ij} = \sum_{k=0}^{m-1} a_{ik} b_{kj} \qquad (5.9)$$

for $i = 0, \ldots, n-1$ and $j = 0, \ldots, p-1$. In the example above, we find c_{21} by summing the products $a_{20}b_{01} + a_{21}b_{11} + a_{22}b_{21}$, which fits Equation 5.9 with $i = 2, j = 1$ and $k = 0, 1, 2$.

Equation 5.9 tells us how to find a single output matrix element. If we loop over i and j, we can find the entire output matrix. This implies a straightforward implementation of matrix multiplication:

```
def matrixmul(A,B):
    I,K = A.shape
    J = B.shape[1]
    C = np.zeros((I,J), dtype=A.dtype)
    for i in range(I):
        for j in range(J):
            for k in range(K):
                C[i,j] += A[i,k]*B[k,j]
    return C
```

We'll assume the arguments, A and B, are compatible matrices. We set the number of rows (I) and columns (J) of the output matrix, C, and use them as the loop limits for the elements of C. We create the output matrix, C, and give it the same data type as A. Then starts a triple loop. The loop over i covers all the rows of the output. The next loop, over j, covers the columns of the current row, and the innermost loop, over k, covers the combining of elements from A and B, as in Equation 5.9. When all loops finish, we return the matrix product, C.

The function matrixmul works. It finds the matrix product. However, in terms of implementation, it's quite naive. Advanced algorithms exist, as do many optimizations of the naive approach when using compiled code. As we'll see below, NumPy supports matrix multiplication and internally uses highly optimized compiled code libraries that far outstrip the performance of the simple code above.

Matrix Notation for Inner and Outer Products

We are now in a position to understand the matrix notation above for the inner product, $a^\top b$, and the outer product, $ab^\top$, of two vectors. In the first case, we have a $1 \times n$ row vector, because of the transpose, and an $n \times 1$ column vector. The algorithm says to form the inner product of the row vector and the column vector to arrive at an output matrix that is 1×1, that is, a single scalar number. Notice that there must be n components in both $\boldsymbol{a}$ and $\boldsymbol{b}$.

For the outer product, we have an $n \times 1$ column vector on the left and a $1 \times m$ row vector on the right. Therefore, we know the output matrix is $n \times m$. If $m = n$, we'll have an output matrix that's $n \times n$. A matrix with as many rows as it has columns is a *square matrix*. These have special properties, some of which we'll see in Chapter 6.

To find the outer product of two vectors by matrix multiplication, we multiply each element of the rows of $\boldsymbol{a}$ by each of the columns of $\boldsymbol{b}$ as a *row vector*,

$$\boldsymbol{ab}^\top = \begin{bmatrix} a_0 \\ a_1 \\ a_2 \end{bmatrix} \begin{bmatrix} b_0 & b_1 & b_2 \end{bmatrix}$$

$$= \begin{bmatrix} a_0 b_0 & a_0 b_1 & a_0 b_2 \\ a_1 b_0 & a_1 b_1 & a_1 b_2 \\ a_2 b_0 & a_2 b_1 & a_2 b_2 \end{bmatrix}$$

where each column of $\boldsymbol{b}^\top$, a single scalar number, is passed down the rows of $\boldsymbol{a}$, thereby forming each possible product between the elements of the two vectors.

We've seen how to perform matrix multiplication manually. Let's take a look now at how NumPy supports matrix multiplication.

Matrix Multiplication in NumPy

NumPy provides two different functions that we can use for matrix multiplication. The first, we've seen already, np.dot, though we've only used it so far to compute inner products of vectors. The second is np.matmul, which is also called when using the @ binary operator available in Python 3.5 and later. Matrix multiplication with either function works as we expect. However, NumPy sometimes treats 1D arrays differently from row or column vectors.

We can use shape to decide if a NumPy array is a 1D array, a row vector, or a column vector, as shown in Listing 5-1:

```
>>> av = np.array([1,2,3])
>>> ar = np.array([[1,2,3]])
>>> ac = np.array([[1],[2],[3]])
>>> av.shape
(3,)
>>> ar.shape
(1, 3)
>>> ac.shape
(3, 1)
```

Listing 5-1: NumPy vectors

Here, we see that a 1D array with three elements, av, has a shape different from a row vector with three components, ar, or a column vector of three components, ac. However, each of these arrays contains the same three integers: 1, 2, and 3.

Let's run an experiment to help us understand how NumPy implements matrix multiplication. We'll test `np.dot`, but the results are the same if we use `np.matmul` or the @ operator. We need a collection of vectors and matrices to work with. We'll then apply combinations of them to `np.dot` and consider the output, which may very well be an error if the operation is undefined for that combination of arguments.

Let's create the arrays, vectors, and matrices we'll need:

```
a1 = np.array([1,2,3])
ar = np.array([[1,2,3]])
ac = np.array([[1],[2],[3]])
b1 = np.array([1,2,3])
br = np.array([[1,2,3]])
bc = np.array([[1],[2],[3]])
A = np.array([[1,2,3],[4,5,6],[7,8,9]])
B = np.array([[9,8,7],[6,5,4],[3,2,1]])
```

The shape of the objects should be discernible from the definition, if we keep the results of Listing 5-1 in mind. We'll also define two 3 × 3 matrices, A and B.

Next, we'll define a helper function to wrap the call to NumPy so we can trap any errors:

```
def dot(a,b):
    try:
        return np.dot(a,b)
    except:
        return "fails"
```

This function calls `np.dot` and returns the word `fails` if the call doesn't succeed. Table 5-1 shows the output of `dot` for the given combinations of the inputs defined above.

Table 5-1 illustrates how NumPy sometimes treats 1D arrays differently from row or column vectors. See the difference in Table 5-1 for a1,A versus ar,A and A,ac. The output of A,ac is what we'd expect to see mathematically, with the column vector a_c multiplied on the left by A.

Is there any real difference between `np.dot` and `np.matmul`? Yes, some. For 1D and 2D arrays, there is no difference. However, there is a difference between how each function handles arrays greater than two dimensions, although we won't work with those here. Also, `np.dot` allows one of its arguments to be a scalar and multiplies each element of the other argument by it. Multiplying by a scalar with `np.matmul` throws an error.

Table 5-1: Results of Applying dot or matmul to Different Types of Arguments

Arguments	Result of np.dot or np.matmul
a1,b1	14 (scalar)
a1,br	fails
a1,bc	[14] (1 vector)
ar,b1	[14] (1 vector)
ar,br	fails
ar,bc	[14] (1 × 1 matrix)
ac,b1	fails
ac,br	$\begin{bmatrix} 1 & 2 & 3 \\ 2 & 4 & 6 \\ 3 & 6 & 9 \end{bmatrix}$ (outer product)
ac,bc	fails
A,a1	[14 32 50] (3 vector)
A,ar	fails
A,ac	$\begin{bmatrix} 14 \\ 32 \\ 50 \end{bmatrix}$
a1,A	[30 36 42] (3 vector)
ar,A	[30 36 42] (1 × 3 matrix)
ac,A	fails
A,B	$\begin{bmatrix} 30 & 24 & 18 \\ 84 & 69 & 54 \\ 138 & 114 & 90 \end{bmatrix}$

Kronecker Product

The final form of matrix multiplication we'll discuss is the *Kronecker product* or *matrix direct product* of two matrices. When computing the matrix product, we mixed individual elements of the matrices, multiplying them together. For the Kronecker product, we multiply the elements of one matrix by an entire matrix to produce an output matrix that is larger than the input matrices. The Kronecker product is also a convenient place to introduce the idea of a *block matrix*, or a matrix constructed from smaller matrices (the blocks).

For example, if we have three matrices

$$A = \begin{bmatrix} 1 & 2 & 3 \\ 4 & 5 & 6 \\ 7 & 8 & 9 \end{bmatrix} \quad B = \begin{bmatrix} 11 & 22 \\ 33 & 44 \\ 55 & 66 \end{bmatrix} \quad C = \begin{bmatrix} 111 \\ 222 \\ 333 \end{bmatrix} \tag{5.10}$$

we can define a block matrix, M, as the following.

$$M = \begin{bmatrix} A & B & C \\ B & C & A \end{bmatrix} = \begin{bmatrix} 1 & 2 & 3 & 11 & 22 & 111 \\ 4 & 5 & 6 & 33 & 44 & 222 \\ 7 & 8 & 9 & 55 & 66 & 333 \\ 11 & 22 & 111 & 1 & 2 & 3 \\ 33 & 44 & 222 & 4 & 5 & 6 \\ 55 & 66 & 333 & 7 & 8 & 9 \end{bmatrix}$$

where each element of M is a smaller matrix stacked on top of each other.

We can most easily define the Kronecker product using a visual example involving a block matrix. The Kronecker product of A and B, typically written as $A \otimes B$, is

$$A \otimes B = \begin{bmatrix} a_{00}B & a_{01}B & \cdots & a_{0,n-1}B \\ a_{10}B & a_{11}B & \cdots & a_{1,n-1}B \\ \vdots & \vdots & & \vdots \\ a_{m-1,0}B & a_{m-1,1}B & \cdots & a_{m-1,n-1}B \end{bmatrix}$$

for A, an $m \times n$ matrix. This is a block matrix because of B, so, when written out completely, the Kronecker product results in a matrix larger than either A or B. Note, unlike matrix multiplication, the Kronecker product is defined for arbitrarily sized A and B matrices. For example, using A and B from Equation 5.10, the Kronecker product is

$$A \otimes B = \begin{bmatrix} (1)B & (2)B & (3)B \\ (4)B & (5)B & (6)B \\ (7)B & (8)B & (9)B \end{bmatrix} = \begin{bmatrix} 11 & 22 & 22 & 44 & 33 & 66 \\ 33 & 44 & 66 & 88 & 99 & 132 \\ 55 & 66 & 110 & 132 & 165 & 198 \\ 44 & 88 & 55 & 110 & 66 & 132 \\ 132 & 176 & 165 & 220 & 198 & 264 \\ 220 & 264 & 275 & 330 & 330 & 396 \\ 77 & 154 & 88 & 176 & 99 & 198 \\ 231 & 308 & 264 & 352 & 297 & 396 \\ 385 & 462 & 440 & 528 & 495 & 594 \end{bmatrix}$$

Notice above that we used $\otimes$ for the Kronecker product. This is the convention, though the symbol $\otimes$ is sometimes abused and is used for other things too. We used it for the outer product of two vectors, for example. NumPy supports the Kronecker product via np.kron.

Summary

In this chapter, we introduced the mathematical objects used in deep learning: scalars, vector, matrices, and tensors. We then explored arithmetic with tensors, in particular with vectors and matrices. We saw how to perform operations on these objects, both mathematically and in code via NumPy.

Our exploration of linear algebra is not complete, however. In the next chapter, we'll dive deeper into matrices and their properties to discuss just a handful of the important things that we can do with or know about them.

6

MORE LINEAR ALGEBRA

In this chapter, we'll continue our exploration of linear algebra concepts. Some of these concepts are only tangentially related to deep learning, but they're the sort of math you'll eventually encounter. Think of this chapter as assumed background knowledge.

Specifically, we'll learn more about the properties of and operations on square matrices, introducing terms you'll encounter in the deep learning literature. After that, I'll introduce the ideas behind the eigenvalues and eigenvectors of a square matrix and how to find them. Next, we'll explore vector norms and other ways of measuring distance that are often encountered in deep learning. At that point, I'll introduce the important concept of a covariance matrix.

We'll conclude the chapter by demonstrating principal component analysis (PCA) and singular value decomposition (SVD). These frequently used approaches depend heavily on the concepts and operators introduced throughout the chapter. We will see what PCA is, how to do it, and what it can buy us from a machine learning perspective. Similarly, we will work with SVD and see how we can use it to implement PCA as well as compute the pseudoinverse of a rectangular matrix.

Square Matrices

Square matrices occupy a special place in the world of linear algebra. Let's explore them in more detail. The terms used here will show up often in deep learning and other areas.

Why Square Matrices?

If we multiply a matrix by a column vector, we'll get another column vector as output:

$$\begin{bmatrix} 1 & 2 & 3 & 4 \\ 5 & 6 & 7 & 8 \end{bmatrix} \begin{bmatrix} 11 \\ 12 \\ 13 \\ 14 \end{bmatrix} = \begin{bmatrix} 130 \\ 330 \end{bmatrix}$$

Interpreted geometrically, the 2×4 matrix has mapped the 4×1 column vector, a point in $\mathbb{R}^4$, to a new point in $\mathbb{R}^2$. The mapping is linear because the point values are only being multiplied by the elements of the 2×4 matrix; there are no nonlinear operations, such as raising the components of the vector to a power, for example.

Viewed this way, we can use a matrix to transform points between spaces. If the matrix is square, say, $n \times n$, the mapping is from $\mathbb{R}^n$ back to $\mathbb{R}^n$. For example, consider

$$\begin{bmatrix} 1 & 2 & 3 \\ 4 & 5 & 6 \\ 7 & 8 & 9 \end{bmatrix} \begin{bmatrix} 11 \\ 12 \\ 13 \end{bmatrix} = \begin{bmatrix} 74 \\ 182 \\ 209 \end{bmatrix}$$

where the point $(11, 12, 13)$ is mapped to the point $(74, 182, 209)$, both in $\mathbb{R}^3$.

Using a matrix to map points from one space to another makes it possible to rotate a set of points about an axis by using a *rotation matrix*. For simple rotations, we can define matrices in 2D,

$$R(\theta) = \begin{bmatrix} \cos\theta & -\sin\theta \\ \sin\theta & \cos\theta \end{bmatrix} \tag{6.1}$$

and in 3D,

$$\begin{bmatrix} 1 & 0 & 0 \\ 0 & \cos\theta & -\sin\theta \\ 0 & \sin\theta & \cos\theta \end{bmatrix}_x \begin{bmatrix} \cos\theta & 0 & \sin\theta \\ 0 & 1 & 0 \\ -\sin\theta & 0 & \cos\theta \end{bmatrix}_y \begin{bmatrix} \cos\theta & -\sin\theta & 0 \\ \sin\theta & \cos\theta & 0 \\ 0 & 0 & 1 \end{bmatrix}_z$$

Rotations are by an angle, θ, and for 3D, about the x-, y-, or z-axis, as indicated by the subscript.

Using a matrix, we can create an *affine transformation*. An affine transformation maps a set of points into another set of points so that points on a line in the original space are still on a line in the mapped space. The transformation is

$$y = Ax + b$$

The affine transform combines a matrix transform, A, with a translation, b, to map a vector, x, to a new vector, y. We can combine this operation into a single matrix multiplication by putting A in the upper-left corner of the matrix and adding b as a new column on the right. A row of all zeros at the bottom with a single 1 in the rightmost column completes the augmented transformation matrix. For an affine transformation matrix

$$\begin{bmatrix} a & b \\ c & d \end{bmatrix}$$

and translation vector

$$\begin{bmatrix} i \\ j \end{bmatrix}$$

we get

$$\begin{bmatrix} x' \\ y' \\ 1 \end{bmatrix} = \begin{bmatrix} a & b & i \\ c & d & j \\ 0 & 0 & 1 \end{bmatrix} \begin{bmatrix} x \\ y \\ 1 \end{bmatrix}$$

This form maps a point, (x, y), to a new point, (x', y').

This maneuver is identical to the *bias trick* sometimes used when implementing a neural network to bury the bias in an augmented weight matrix by including an extra feature vector input set to 1. In fact, we can view a feedforward neural network as a series of affine transformations, where the transformation matrix is the weight matrix between the layers, and the bias vector provides the translation. The activation function at each layer alters the otherwise linear relationship between the layers. It is this nonlinearity that lets the network learn a new way to map inputs so that the final output reflects the functional relationship the network is designed to learn.

We use square matrices, then, to map points from one space back into the same space, for example to rotate them about an axis. Let's look now at some special properties of square matrices.

Transpose, Trace, and Powers

Chapter 5 showed us the vector transpose to move between column and row vectors. The transpose operation is not restricted to vectors. It works for any matrix by flipping the rows and columns along the main diagonal. For example,

$$\begin{bmatrix} 1 & 2 & 3 \\ 4 & 5 & 6 \\ 7 & 8 & 9 \end{bmatrix}^\top = \begin{bmatrix} 1 & 4 & 7 \\ 2 & 5 & 8 \\ 3 & 6 & 9 \end{bmatrix}, \begin{bmatrix} a & b & c & d \\ e & f & g & h \end{bmatrix}^\top = \begin{bmatrix} a & e \\ b & f \\ c & g \\ d & h \end{bmatrix}$$

The transpose is formed by flipping the indices of the matrix elements:

$$a_{ji} \leftarrow a_{ij}, \; i = 0, 1, \ldots, n-1, \; j = 0, 1, \ldots, m-1$$

This changes an $n \times m$ matrix into an $m \times n$ matrix. Notice that the order of a square matrix remains the same under the transpose operation, and the values on the main diagonal don't change.

In NumPy, you can call the transpose method on an array, but the transpose is so common that a shorthand notation (.T) also exists. For example,

```
>>> import numpy as np
>>> a = np.array([[1,2,3],[4,5,6],[7,8,9]])
>>> print(a)
[[1 2 3]
 [4 5 6]
 [7 8 9]]
>>> print(a.transpose())
[[1 4 7]
 [2 5 8]
 [3 6 9]]
>>> print(a.T)
[[1 4 7]
 [2 5 8]
 [3 6 9]]
```

The *trace* is another common operation applied to square matrices:

$$\mathrm{tr}\boldsymbol{A} = \sum_{i=0}^{n-1} a_{ii}$$

As an operator, the trace has certain properties. For example, it's linear:

$$\mathrm{tr}(\boldsymbol{A} + \boldsymbol{B}) = \mathrm{tr}\boldsymbol{A} + \mathrm{tr}\boldsymbol{B}$$

It's also true that $\mathrm{tr}(\boldsymbol{A}) = \mathrm{tr}(\boldsymbol{A}^T)$ and $\mathrm{tr}(\boldsymbol{AB}) = \mathrm{tr}(\boldsymbol{BA})$.

NumPy uses np.trace to quickly calculate the trace of a matrix and np.diag to return the diagonal elements of a matrix as a 1D array,

$$(a_{00}, a_{11}, \ldots, a_{n-1,n-1})$$

for an $n \times n$ or $n \times m$ matrix.

A matrix doesn't need to be square for NumPy to return the elements along its diagonal. And although mathematically the trace generally only applies to square matrices, NumPy will calculate the trace of any matrix, returning the sum of the diagonal elements:

```
>>> b = np.array([[1,2,3,4],[5,6,7,8]])
>>> print(b)
```

```
[[1 2 3 4]
 [5 6 7 8]]
>>> print(np.diag(b))
[1 6]
>>> print(np.trace(b))
7
```

Lastly, you can multiply a square matrix by itself, implying that you can raise a square matrix to an integer power, n, by multiplying itself n times. Note that this is not the same as raising the elements of the matrix to a power. For example,

$$A = \begin{bmatrix} 1 & 2 \\ 3 & 4 \end{bmatrix}, \; A^2 = AA = \begin{bmatrix} 1 & 2 \\ 3 & 4 \end{bmatrix} \begin{bmatrix} 1 & 2 \\ 3 & 4 \end{bmatrix} = \begin{bmatrix} 7 & 10 \\ 15 & 22 \end{bmatrix}$$
$$\neq \begin{bmatrix} 1 & 4 \\ 9 & 16 \end{bmatrix}$$

The matrix power follows the same rules as raising any number to a power:

$$A^n A^m = A^{n+m}$$

$$(A^n)^m = A^{nm}$$

for $n, m \in \mathbb{I}^+$ (positive integers) and where A is a square matrix.

NumPy provides a function to compute the power of a square matrix more efficiently than repeated calls to np.dot:

```
>>> from numpy.linalg import matrix_power
>>> a = np.array([[1,2],[3,4]])
>>> print(matrix_power(a,2))
[[ 7 10]
 [15 22]]
>>> print(matrix_power(a,10))
[[ 4783807  6972050]
 [10458075 15241882]]
```

Now let's consider some special square matrices that you'll encounter from time to time.

Special Square Matrices

Many square (and nonsquare) matrices have received special names. Some are rather obvious, like matrices that are all zero or one, which are called *zeros matrices* and *ones matrices*, respectively. NumPy uses these extensively:

```
>>> print(np.zeros((3,5)))
[[0. 0. 0. 0. 0.]
 [0. 0. 0. 0. 0.]
```

```
[0. 0. 0. 0. 0.]]
>>> print(np.ones(3,3))
[[1. 1. 1.]
 [1. 1. 1.]
 [1. 1. 1.]]
```

Note that you can find a matrix of any constant value, c, by multiplying the ones matrix by c.

Notice above that NumPy defaults to matrices of 64-bit floating-point numbers corresponding to a C-language type of `double`. See Table 1-1 on page 6 for a list of possible numeric data types. You can specify the desired data type with the `dtype` keyword. In pure mathematics, we don't care much about data types, but to work in deep learning, you need to pay attention to avoid defining arrays that are far more memory-hungry than needed. Many deep learning models are happy with arrays of 32-bit floats, which use half the memory per element than the NumPy default. Also, many toolkits make use of new or previously seldom-used data types, like 16-bit floats, to allow for even better use of memory. NumPy does support 16-bit floats by specifying `float16` as the `dtype`.

The Identity Matrix

By far, the most important special matrix is the *identity matrix*. This is a square matrix with all ones on the diagonal:

$$I = \begin{bmatrix} 1 & 0 & 0 & \cdots & 0 \\ 0 & 1 & 0 & \cdots & 0 \\ 0 & 0 & 1 & \cdots & 0 \\ \vdots & \vdots & \vdots & & \vdots \\ 0 & 0 & 0 & \cdots & 1 \end{bmatrix} \tag{6.2}$$

The identity matrix acts like the number 1 when multiplying a matrix. Therefore,

$$AI = IA = A$$

for an $n \times n$ square matrix A and an $n \times n$ identity matrix I. When necessary, we'll add a subscript to indicate the order of the identity matrix, for example, I_n.

NumPy uses `np.identity` or `np.eye` to generate identity matrices of a given size:

```
>>> a = np.array([[1,2],[3,4]])
>>> i = np.identity(2)
>>> print(i)
[[1. 0.]
 [0. 1.]]
>>> print(a @ i)
```

```
[[1. 2.]
 [3. 4.]]
```

Look carefully at the example above. Mathematically, we said that multiplication of a square matrix by the identity matrix of the same order returns the matrix. NumPy, however, did something we might not want. Matrix a was defined with integer elements, so it has a data type of `int64`, the NumPy default for integers. However, since we didn't explicitly provide `np.identity` with a data type, NumPy defaulted to a 64-bit float. Therefore, matrix multiplication (`@`) between a and i returned a floating-point version of a. This subtle change of data type might be important for later calculations, so, again, we need to pay attention to data types when using NumPy.

It doesn't matter if you use `np.identity` or `np.eye`. In fact, internally, `np.identity` is just a wrapper for `np.eye`.

Triangular Matrices

Occasionally, you'll hear about *triangular* matrices. There are two kinds: upper and lower. As you may intuit from the name, an upper triangular matrix is one with nonzero elements in the part on or above the main diagonal, whereas a lower triangular matrix only has elements on or below the main diagonal. For example,

$$U = \begin{bmatrix} 1 & 2 & 3 & 4 \\ 0 & 5 & 6 & 7 \\ 0 & 0 & 8 & 9 \\ 0 & 0 & 0 & 10 \end{bmatrix}$$

is an upper triangular matrix, whereas

$$L = \begin{bmatrix} 1 & 0 & 0 & 0 \\ 2 & 3 & 0 & 0 \\ 4 & 5 & 6 & 0 \\ 7 & 8 & 9 & 10 \end{bmatrix}$$

is a lower triangular matrix. A matrix that has elements only on the main diagonal is, not surprisingly, a *diagonal matrix*.

NumPy has two functions, `np.triu` and `np.tril`, to return the upper or lower triangular part of the given matrix, respectively. So,

```
>>> a = np.arange(16).reshape((4,4))
>>> print(a)
[[ 0  1  2  3]
 [ 4  5  6  7]
 [ 8  9 10 11]
 [12 13 14 15]]
>>> print(np.triu(a))
[[ 0  1  2  3]
 [ 0  5  6  7]
 [ 0  0 10 11]
```

```
 [ 0  0  0 15]]
>>> print(np.tril(a))
[[ 0  0  0  0]
 [ 4  5  0  0]
 [ 8  9 10  0]
 [12 13 14 15]]
```

We don't frequently use triangular matrices in deep learning, but we do use them in linear algebra, in part to compute determinants, to which we now turn.

Determinants

We can think of the *determinant* of a square matrix, $n \times n$, as a function mapping square matrices to a scalar. The primary use of the determinant in deep learning is to compute the eigenvalues of a matrix. We'll see what that means later in this chapter, but for now think of eigenvalues as special scalar values associated with a matrix. The determinant also tells us something about whether or not a matrix has an inverse, as we'll also see below. Notationally, we write the determinant of a matrix with vertical bars. For example, if A is a 3×3 matrix, we write the determinant as

$$\det(A) = \begin{vmatrix} a_{00} & a_{01} & a_{02} \\ a_{10} & a_{11} & a_{12} \\ a_{20} & a_{21} & a_{22} \end{vmatrix} \in \mathbb{R}$$

where we state explicitly that the value of the determinant is a scalar (element of $\mathbb{R}$). All square matrices have a determinant. For now, let's consider some of the properties of the determinant:

1. If any row or column of A is zero, then $\det(A) = 0$.

2. If any two rows of A are identical, then $\det(A) = 0$.

3. If A is an upper or lower triangular, then $\det(A) = \prod_{i=0}^{n-1} a_{ii}$.

4. If A is a diagonal matrix, then $\det(A) = \prod_{i=0}^{n-1} a_{ii}$.

5. The determinant of the identity matrix, regardless of size, is 1.

6. The determinant of a product of matrices is the product of the determinants, $\det(AB) = \det(A)\det(B)$.

7. $\det(A) = \det(A^\top)$.

8. $\det(A^n) = \det(A)^n$.

Property 7 indicates that the transpose operation does not change the value of a determinant. Property 8 is a consequence of Property 6.

We have multiple ways we can calculate the determinant of a square matrix. We'll examine only one way here, which involves using a recursive formula. All recursive formulas apply themselves, just as recursive functions in code call themselves. The general idea is that each recursion works on a simpler version of the problem, which can be combined to return the solution to the larger problem.

For example, we can calculate the factorial of an integer,

$$n! = n(n-1)(n-2)(n-3)\ldots 1$$

recursively if we notice the following:

$$n! = n \times (n-1)!$$

$$0! = 1$$

The first statement says that the factorial of n is n times the factorial of $(n-1)$. The second statement says that the factorial of zero is one. The recursion is the first statement, but this recursion will never end without some condition that returns a value. That's the point of the second statement, the *base case*: it says the recursion ends when we get to zero.

This might be clearer in code. We can define the factorial like so:

```
def factorial(n):
    if (n == 0):
        return 1
    return n*factorial(n-1)
```

Notice that `factorial` calls itself on the argument minus one, unless the argument is zero, in which case it immediately returns one. The code works because of the Python call stack. The call stack keeps track of all the computations of `n*factorial(n-1)`. When we encounter the base case, all the pending multiplications are done, and we return the final value.

To calculate determinants recursively, then, we need a recursion statement, something that defines the determinant of a matrix in terms of simpler determinants. We also need a base case that gives us a definitive value. For determinants, the base case is when we get to a 1×1 matrix. For any 1×1 matrix, A, we have

$$\det(A) = a_{00}$$

meaning the determinant of a 1×1 matrix is the single value it contains.

Our plan is to calculate the determinant by breaking the calculation into successively simpler determinants until we reach the base case above. To do this, we need a statement involving recursion. However, we need to define a few things before we can make the statement. First, we need to define the *minor* of a matrix. The (i,j)-minor of a matrix, A, is the matrix left after removing the ith row and jth column of A. We'll denote a minor matrix by A_{ij}. For example, given

$$A = \begin{bmatrix} 9 & 8 & 7 \\ 6 & 5 & 4 \\ 3 & 2 & 1 \end{bmatrix}$$

then

$$A_{11} = \begin{bmatrix} \underline{\textbf{9}} & 8 & \underline{\textbf{7}} \\ 6 & 5 & 4 \\ \underline{\textbf{3}} & 2 & \underline{\textbf{1}} \end{bmatrix} = \begin{bmatrix} 9 & 7 \\ 3 & 1 \end{bmatrix}$$

where the minor, A_{11}, is found by deleting row 1 and column 1 to leave only the underlined values.

Second, we need to define the *cofactor*, C_{ij}, of the minor, A_{ij}. This is where our recursive statement appears. The definition is

$$C_{ij} = (-1)^{i+j+2}\det(A_{ij})$$

The cofactor depends on the determinant of the minor. Notice the exponent on -1, written as $i + j + 2$. If you look at most math books, you'll see the exponent as $i + j$. We've made a conscious choice to define matrices with zero-based indices so the math and implementation in code match without being off by one. Here's one place where that choice forces us to be less elegant than the math texts. Because our indices are "off" by one, we need to add that one back into the exponent of the cofactor so the pattern of positive and negative values that the cofactor uses is correct. This means adding one to each of the variables in the exponent: $i \rightarrow i + 1$ and $j \rightarrow j + 1$. This makes the exponent $i + j \rightarrow (i + 1) + (j + 1) = i + j + 2$.

We're now ready for our full recursive definition of the determinant of A by using *cofactor expansion*. It turns out that summing the product of the matrix values and associated cofactors for any row or column of a square matrix will give us the determinant. So, we'll use the first row of the matrix and calculate the determinant as

$$\det(A) = \sum_{j=0}^{n-1} a_{0j} C_{0j} \tag{6.3}$$

You may be wondering: Where's the recursion in Equation 6.3? It shows up on the determinant of the minor. If A is an $n \times n$ matrix, the minor, A_{ij}, is an $(n - 1) \times (n - 1)$ matrix. Therefore, to calculate the cofactors to find the determinant of an $n \times n$ matrix, we need to know how to find the determinant of an $(n - 1) \times (n - 1)$ matrix. However, we can use cofactor expansion to find the $(n - 1) \times (n - 1)$ determinant, which involves finding the determinant of an $(n - 2) \times (n - 2)$ matrix. This process continues until we get to a 1×1 matrix. We already know the determinant of a 1×1 matrix is the single value it contains.

Let's work through this process for a 2×2 matrix:

$$A = \begin{bmatrix} a & b \\ c & d \end{bmatrix}$$

Using cofactor expansion, we get

$$\det(A) = a_{00}C_{00} + a_{01}C_{01}$$

$$= aC_{00} + bC_{01}$$

$$= a(-1)^{0+0+2}\det(A_{00}) + b(-1)^{0+1+2}\det(A_{01})$$

$$= ad - bc$$

which is the formula for the determinant of a 2×2 matrix. The minors of a 2×2 matrix are 1×1 matrices, each returning either d or c in this case.

In NumPy, we calculate determinants with $\texttt{np.linalg.det}$. For example,

```
>>> a = np.array([[1,2],[3,4]])
>>> print(a)
[[1 2]
 [3 4]]
>>> np.linalg.det(a)
-2.0000000000000004
>>> 1*4 - 2*3
-2
```

The last line of code uses the formula for a 2×2 matrix we derived above for comparison purposes. Internally, NumPy does not use recursive cofactor expansion to calculate the determinant. Instead, it factors the matrix into the product of three matrices: (1) a *permutation matrix*, which looks like a scrambled identity matrix with only a single one in each row and column, (2) a lower triangular matrix, and (3) an upper triangular matrix. The determinant of the permutation matrix is either $+1$ or -1. The determinant of a triangular matrix is the product of the diagonal elements, while the determinant of a product of matrices is the product of the per-matrix determinants.

We can use determinants to determine whether a matrix has an inverse. Let's turn there now.

Inverses

Equation 6.2 defines the identity matrix. We said that this matrix acts like the number 1, so when it multiplies a square matrix, the same square matrix is returned. When multiplying scalars, we know that for any number, $x \neq 0$, there exists another number, call it y, such that $xy = 1$. This number is the multiplicative inverse of x. Furthermore, we know exactly what y is; it's $1/x = x^{-1}$.

By analogy, then, we might wonder if, since we have an identity matrix that acts like the number 1, there is another square matrix, call it A^{-1}, for a given square matrix, A, such that

$$AA^{-1} = A^{-1}A = I$$

If A^{-1} exists, it's known as the *inverse matrix* of A, and A is said to be *invertable*. For real numbers, all numbers except zero have an inverse. For matrices, it isn't so straightforward. Many square matrices don't have inverses. To check if A has an inverse, we use the determinant: $\det(A) = 0$ tells us that A has no inverse. Furthermore, if A^{-1} exists, then

$$\det(A^{-1}) = \frac{1}{\det(A)}$$

Note also that $(A^{-1})^{-1} = A$, as is the case for real numbers. Another useful property of inverses is

$$(AB)^{-1} = B^{-1}A^{-1}$$

where the order of the product on the right-hand side is important. Finally, note that the inverse of a diagonal matrix is simply the reciprocal of the diagonal elements:

$$A = \begin{bmatrix} a & 0 & 0 \\ 0 & b & 0 \\ 0 & 0 & c \end{bmatrix} \rightarrow A^{-1} = \begin{bmatrix} a^{-1} & 0 & 0 \\ 0 & b^{-1} & 0 \\ 0 & 0 & c^{-1} \end{bmatrix}$$

It's possible to calculate the inverse by hand using row operations, which we've conveniently ignored here because they are seldom used in deep learning. Cofactor expansion techniques can also calculate the inverse, but to save time, we won't elaborate on the process here. What's important for us is to know that square matrices often have an inverse, and that we can calculate inverses with NumPy via np.linalg.inv.

If a matrix is *not* invertible, the matrix is said to be *singular*. Therefore, the determinant of a singular matrix is zero. If a matrix has an inverse, it is a *nonsingular* or *nondegenerate* matrix.

In NumPy, we use np.linalg.inv to calculate the inverse of a square matrix. For example,

```
>>> a = np.array([[1,2,1],[2,1,2],[1,2,2]])
>>> print(a)
[[1 2 1]
 [2 1 2]
 [1 2 2]]
>>> b = np.linalg.inv(a)
>>> print(b)
[[ 0.66666667  0.66666667 -1. ]
```

```
[ 0.66666667 -0.33333333  0. ]
[-1.          0.          1. ]]
>>> print(a @ b)
[[1. 0. 0.]
 [0. 1. 0.]
 [0. 0. 1.]]
>>> print(b @ a)
[[1. 0. 0.]
 [0. 1. 0.]
 [0. 0. 1.]]
```

Notice the inverse (b) working as we expect and giving the identity matrix when multiplying a from the left or right.

Symmetric, Orthogonal, and Unitary Matrices

If for a square matrix, A, we have

$$A^\top = A$$

then A is said to be a *symmetric matrix*. For example,

$$A = \begin{bmatrix} 1 & 2 & 3 & 4 \\ 2 & 5 & 6 & 7 \\ 3 & 6 & 8 & 9 \\ 4 & 7 & 9 & 1 \end{bmatrix}$$

is a symmetric matrix, since $A^\top = A$.

Notice that diagonal matrices are symmetric, and the product of two symmetric matrices is commutative: $AB = BA$. The inverse of a symmetric matrix, if it exists, is also a symmetric matrix.

If the following is true,

$$AA^\top = A^\top A = I$$

then A is an orthogonal matrix. If A is an orthogonal matrix, then

$$A^{-1} = A^\top$$

and, as a result,

$$\det(A) = \pm 1$$

If the values in the matrix are allowed to be complex, which does not happen often in deep learning, and

$$U^* U = UU^* = I$$

then U is a *unitary matrix* with U^* being the *conjugate transpose* of U. The conjugate transpose is the ordinary matrix transpose followed by the complex conjugate operation to change $i = \sqrt{-1}$ to $-i$. So, we might have

$$U = \begin{bmatrix} 1 + 3i & 2 - 3i \\ 3 + 2i & 3 + 2i \end{bmatrix} \rightarrow U^* = \begin{bmatrix} 1 - 3i & 3 - 2i \\ 2 + 3i & 3 - 2i \end{bmatrix}$$

Sometimes, especially in physics, the conjugate transpose is called the *Hermitian adjoint* and is denoted as $A^{\dagger}$. If a matrix is equal to its conjugate transpose, it is called a *Hermitian matrix*. Notice that real symmetric matrices are also Hermitian matrices because the conjugate transpose is the same as the ordinary transpose when the values are real numbers. Therefore, you might encounter the term *Hermitian* in place of *symmetric* when referring to matrices with real elements.

Definiteness of a Symmetric Matrix

We saw at the beginning of this section that an $n \times n$ square matrix maps a vector in $\mathbb{R}^n$ to another vector in $\mathbb{R}^n$. Let's consider now a symmetric $n \times n$ matrix, B, with real-valued elements. We can characterize this matrix by how it maps vectors using the inner product between the mapped vector and the original vector. Specifically, if x is a column vector ($n \times 1$), then Bx is also an $n \times 1$ column vector. Therefore, the inner product of this vector and the original vector, x, is $x^{\top} Bx$, a scalar.

If the following is true:

$$x^{\top} Bx > 0, \ \forall x \neq 0$$

then B is said to be *positive definite*. Here, the bolded 0 is the $n \times 1$ column vector of all zeros, and $\forall$ is math notation meaning "for all."

Similarly, if

$$x^{\top} Bx < 0, \ \forall x \neq 0$$

then B is *negative definite*. Relaxing the inner product relationship and the nonzero requirement on x gives two additional cases. If

$$x^{\top} Bx \geq 0, \ \forall x \in \mathbb{R}^{n \times 1}$$

then B is said to be *positive semidefinite*, and

$$x^{\top} Bx \leq 0, \ \forall x \in \mathbb{R}^{n \times 1}$$

makes B a *negative semidefinite* matrix. Finally, a real square symmetric matrix that is neither positive nor negative semidefinite is called an *indefinite matrix*. The definiteness of a matrix tells us something about the eigenvalues, which we'll learn more about in the next section. If a symmetric matrix is positive definite, then all of its eigenvalues are positive. Similarly, a symmetric negative definite matrix has all negative eigenvalues. Positive and negative semidefinite symmetric matrices have eigenvalues that are all positive or zero or all negative or zero, respectively.

Let's shift gears now from talking about types of matrices to discovering the importance of eigenvectors and eigenvalues, key properties of a matrix that we use frequently in deep learning.

Eigenvectors and Eigenvalues

We learned above that a square matrix maps a vector into another vector in the same dimensional space, $v' = Av$, where both v' and v are n-dimensional vectors, if A is an $n \times n$ matrix.

Consider this equation,

$$Av = \lambda v \tag{6.4}$$

for some square matrix, A, where λ is a scalar value and v is a nonzero column vector.

Equation 6.4 says that the vector, v, is mapped by A back into a scalar multiple of itself. We call v an *eigenvector* of A with *eigenvalue* λ. The prefix *eigen* comes from German and is often translated as "self," "characteristic," or even "proper." Thinking geometrically, Equation 6.4 says that the action of A on its eigenvectors in $\mathbb{R}^n$ is to shrink or expand the vector without changing its direction. Note, while v is nonzero, it's possible for λ to be zero.

How does Equation 6.4 relate to the identity matrix, I? By definition, the identity matrix maps a vector back into itself without scaling it. Therefore, the identity matrix has an infinite number of eigenvectors, and all of them have an eigenvalue of 1, since, for any x, $Ix = x$. Therefore, the same eigenvalue may apply to more than one eigenvector.

Recall that Equation 6.1 defines a rotation matrix in 2D space for some given angle, θ. This matrix has no eigenvectors, because, for any nonzero vector, it rotates the vector by θ, so it can never map a vector back into its original direction. Therefore, not every matrix has eigenvectors.

Finding Eigenvalues and Eigenvectors

To find the eigenvalues of a matrix, if there are any, we go back to Equation 6.4 and rewrite it:

$$Av - \lambda v = Av - \lambda Iv = (A - \lambda I)v = 0 \tag{6.5}$$

We can insert the identity matrix, I, between λ and v because $Iv = v$. Therefore, to find the eigenvalues of A, we need to find values of λ that cause the matrix $A - \lambda I$ to map a nonzero vector, v, to the zero vector. Equation 6.5 only has solutions other than the zero vector if the determinant of $A - \lambda I$ is zero.

The above gives us a way to find the eigenvalues. For example, consider what $A - \lambda I$ looks like for a 2×2 matrix:

$$A - \lambda I = \begin{bmatrix} a & b \\ c & d \end{bmatrix} - \lambda \begin{bmatrix} 1 & 0 \\ 0 & 1 \end{bmatrix}$$

$$= \begin{bmatrix} a & b \\ c & d \end{bmatrix} - \begin{bmatrix} \lambda & 0 \\ 0 & \lambda \end{bmatrix}$$

$$= \begin{bmatrix} a - \lambda & b \\ c & d - \lambda \end{bmatrix}$$

We learned above that the determinant of a 2×2 matrix has a simple form; therefore, the determinant of the matrix above is

$$\det(A - \lambda I) = (a - \lambda)(d - \lambda) - bc$$

This equation is a second-degree polynomial in λ. Since we need the determinant to be zero, we set this polynomial to zero and find the roots. The roots are the eigenvalues of A. The polynomial that this process finds is called the *characteristic polynomial*, and Equation 6.5 is the *characteristic equation*. Notice above that the characteristic polynomial is a second-degree polynomial. In general, the characteristic polynomial of an $n \times n$ matrix is of degree n, so a matrix has at most n distinct eigenvalues, since an nth degree polynomial has at most n roots.

Once we know the roots of the characteristic polynomial, we can go back to Equation 6.5, substitute each root for λ, and solve to find the associated eigenvectors, the v's of Equation 6.5.

The eigenvalues of a triangular matrix, which includes diagonal matrices, are straightforward to calculate because the determinant of such a matrix is simply the product of the main diagonal. For example, for a 4×4 triangular matrix, the determinant of the characteristic equation is

$$\det(A - \lambda I) = (a_{00} - \lambda)(a_{11} - \lambda)(a_{22} - \lambda)(a_{33} - \lambda)$$

which has four roots: the values of the diagonal. For triangular and diagonal matrices, the entries on the main diagonal *are* the eigenvalues.

Let's see a worked eigenvalue example for the following matrix:

$$A = \begin{bmatrix} 0 & 1 \\ -2 & -3 \end{bmatrix}$$

I selected this matrix to make the math nicer, but the process works for any matrix. The characteristic equation means we need the λ values that make the determinant zero, as shown next.

$$\begin{vmatrix} -\lambda & 1 \\ -2 & -3-\lambda \end{vmatrix} = (-\lambda)(-3-\lambda) + 2$$

$$= \lambda^2 + 3\lambda + 2$$

$$= (\lambda+1)(\lambda+2)$$

The characteristic polynomial is easily factored to give $\lambda = -1, -2$.

In code, to find the eigenvalues and eigenvectors of a matrix, we use np.linalg.eig. Let's check our calculation above to see if NumPy agrees:

```
>>> a = np.array([[0,1],[-2,-3]])
>>> print(np.linalg.eig(a)[0])
[-1. -2.]
```

The np.linalg.eig function returns a list. The first element is a vector of the eigenvalues of the matrix. The second element, which we are ignoring for the moment, is a matrix, the *columns* of which are the eigenvectors associated with each of the eigenvalues. Note that we also could have used np.linalg.eigvals to return just the eigenvalues. Regardless, we see that our calculation of the eigenvalues of A is correct.

To find the associated eigenvectors, we put each of the eigenvalues back into Equation 6.5 and solve for v. For example, for $\lambda = -1$, we get

$$(A - (-1)I)v = 0$$

$$\begin{bmatrix} 1 & 1 \\ -2 & -2 \end{bmatrix} \begin{bmatrix} v_0 \\ v_1 \end{bmatrix} = \begin{bmatrix} 0 \\ 0 \end{bmatrix}$$

which leads to the system of equations:

$$v_0 + v_1 = 0$$

$$-2v_0 - 2v_1 = 0$$

This system has many solutions, as long as $v_0 = -v_1$. That means we can pick v_0 and v_1, as long as the relationship between them is preserved. Therefore, we have our eigenvector for

$$\lambda = -1, v_0 = \begin{bmatrix} 1 \\ -1 \end{bmatrix}$$

If we repeat this process for $\lambda = -2$, we get the relationship between the components of v_2 to be $2v_0 = -v_1$. Therefore, we select $v_1 = \begin{bmatrix} -1 \\ 2 \end{bmatrix}$ as the second eigenvector.

Let's see if NumPy agrees with us. This time, we'll display the second list element returned by np.linalg.eig. This is a matrix where the columns of the matrix are the eigenvectors:

```
>>> print(np.linalg.eig(a)[1])
[[ 0.70710678 -0.4472136 ]
 [-0.70710678  0.89442719]]
```

Hmm... the columns of this matrix do not appear to match our selected eigenvectors. But don't worry—we didn't make a mistake. Recall that the eigenvectors were not uniquely determined, only the relationship between the components was determined. If we'd wanted to, we could have selected other values, as long as for one eigenvector they were of equal magnitude and opposite sign, and for the other they were in the ratio of 2:1 with opposite signs. What NumPy returns is a set of eigenvectors that are of unit length. So, to see that our hand calculation is correct, we need to make our eigenvectors unit vectors by dividing each component by the square root of the sum of the squares of the components. In code, it's succinct, if a bit messy:

```
>>> np.array([1,-1])/np.sqrt((np.array([1,-1])**2).sum())
array([ 0.70710678, -0.70710678])
>>> np.array([-1,2])/np.sqrt((np.array([-1,2])**2).sum())
array([-0.4472136 , 0.89442719])
```

Now we see that we're correct. The unit vector versions of the eigenvectors do match the columns of the matrix NumPy returned.

We'll use the eigenvectors and eigenvalues of a matrix often when we're doing deep learning. For example, we'll see them again later in the chapter when we investigate principal component analysis. But before we can learn about PCA, we need to change focus once again and learn about vector norms and distance metrics commonly used in deep learning, especially about the covariance matrix.

Vector Norms and Distance Metrics

In common deep learning parlance, people are somewhat sloppy and use the terms *norm* and *distance* interchangeably. We can forgive them for doing so; the difference between the terms is small in practice, as we'll see below.

A *vector norm* is a function that maps a vector, real or complex, to some value, $x \in \mathbb{R}$, $x \geq 0$. A norm must satisfy some specific properties in a mathematical sense, but in practice, not everything that's called a norm is, in fact, a norm. In deep learning, we usually use norms as distances between pairs of vectors. In practice, an important property for a distance measure is that the order of the inputs doesn't matter. If $f(x, y)$ is a distance, then $f(x, y) = f(y, x)$. Again, this is not rigorously followed; for example, you'll often see the Kullback-Leibler divergence (KL-divergence) used as a distance even though this property doesn't hold.

Let's start with vector norms and see how we can easily use them as a distance measure between vectors. Then we'll introduce the important concept of a covariance matrix, heavily used on its own in deep learning, and see how we can create a distance measure from it: the Mahalanobis distance. We'll end the section by introducing the KL-divergence, which we can view as a measure between two discrete probability distributions.

L-Norms and Distance Metrics

For an n-dimensional vector, x, we define the p-norm of the vector to be

$$\|x\|_p \equiv \left(\sum_i |x_i|^p \right)^{\frac{1}{p}} \tag{6.6}$$

where p is a real number. Although we use p in the definition, people generally refer to these as L_p norms. We saw one of these norms in Chapter 5 when we defined the magnitude of a vector. In that case, we were calculating the *L2-norm*,

$$\|x\|_2 = \sqrt{x_0^2 + x_1^2 + x_2^2 + \cdots x_{n-1}^2} = \sqrt{x^\top x}$$

which is the square root of the inner product of x with itself.

The norms we use most often in deep learning are the L2-norm and the *L1-norm*,

$$\|x\|_1 = \sum_i |x_i|$$

which is nothing more than the sum of the absolute values of the components of x. Another norm you'll encounter is the L_∞*-norm*,

$$L_\infty = \max |x_i|$$

the maximum absolute value of the components of x.

If we replace x with the difference of two vectors, $x - y$, we can treat the norms as distance measures between the two vectors. Alternatively, we can picture the process as computing the vector norm on the vector that is the difference between x and y.

Switching from norm to distance makes a trivial change in Equation 6.6:

$$L_p(x, y) = \left(\sum_i |x_i - y_i|^p \right)^{\frac{1}{p}} \tag{6.7}$$

The L2-distance becomes

$$L_2 = \sqrt{(x_0 - y_0)^2 + (x_1 - y_1)^2 + (x_2 - y_2)^2 + \cdots (x_{n-1} - y_{n-1})^2}$$

This is the *Euclidean distance* between two vectors. The L1-distance is often called the *Manhattan distance* (also called *city block distance*, *boxcar distance*, or *taxicab distance*):

$$L_1 = \sum_i |x_i - y_i|$$

It's so named because it corresponds to the length a taxicab would travel on the grid of streets in Manhattan. The L_∞-distance is sometimes known as the *Chebyshev distance*.

Norm equations have other uses in deep learning. For example, weight decay, used in deep learning as a regularizer, uses the L2-norm of the weights of the model to keep the weights from getting too large. The L1-norm of the weights is also sometimes used as a regularizer.

Let's move now to consider the important concept of a covariance matrix. It isn't a distance metric itself but is used by one, and it will show up again later in the chapter.

Covariance Matrices

If we have a collection of measurements on multiple variables, like a training set with feature vectors, we can calculate the variance of the features with respect to each other. For example, here's a matrix of observations of four variables, one per row:

$$X = \begin{bmatrix} 5.1 & 3.5 & 1.4 & 0.2 \\ 4.9 & 3.0 & 1.4 & 0.2 \\ 4.7 & 3.2 & 1.3 & 0.2 \\ 4.6 & 3.1 & 1.5 & 0.2 \\ 5.0 & 3.6 & 1.4 & 0.2 \end{bmatrix}$$

In reality, X is the first five samples from the famous iris dataset. For the iris dataset, the features are measurements of the parts of iris flowers from three different species. You can load this dataset into NumPy using sklearn:

```
>>> from sklearn import datasets
>>> iris = datasets.load_iris()
>>> X = iris.data[:5]
>>> X
array([[5.1, 3.5, 1.4, 0.2],
       [4.9, 3.0, 1.4, 0.2],
       [4.7, 3.2, 1.3, 0.2],
       [4.6, 3.1, 1.5, 0.2],
       [5.0, 3.6, 1.4, 0.2]])
```

We could calculate the standard deviation of each of the features, the columns of A, but that would only tell us about the variance of the values of that feature around its mean. Since we have multiple features, it would be nice to know something about how the features in, say, column zero and column one vary together. To determine this, we need to calculate the *co-variance matrix*. This matrix captures the variance of the individual features along the main diagonal. Meanwhile, the off-diagonal values represent how one feature varies as another varies—these are the covariances. Since there are four features, the covariance matrix, which is always square, is, in this case, a 4×4 matrix. We find the elements of the covariance matrix, Σ, by calculating

$$\Sigma_{ij} = \frac{1}{n-1} \sum_{k=0}^{n-1} (x_{ki} - \bar{x}_i)(x_{kj} - \bar{x}_j) \tag{6.8}$$

assuming the rows of the matrix, X, are the observations, and the columns of X represent the different features. The means of the features across all rows are $\bar{x}_i$ and $\bar{x}_j$ for features i and j. Here, n is the number of observations, the number of rows in X. We can see that when $i = j$, the covariance value is the normal variance for that feature. When $i \neq j$, the value is how i and j vary together. We often denote the covariance matrix as Σ, and it is always symmetric: $\Sigma_{ij} = \Sigma_{ji}$.

Let's calculate some elements of the covariance matrix for X above. The per-feature means are $\bar{x} = [4.86, 3.28, 1.4, 0.2]$. Let's find the first row of Σ. This will tell us the variance of the first feature (column of X) and how that feature varies with the second, third, and fourth features. Therefore, we need to calculate Σ_{00}, Σ_{01}, Σ_{02}, and Σ_{03}:

$$\Sigma_{00} = \frac{1}{5-1} \sum_{k=0}^{4} (x_{k0} - 4.86)(x_{k0} - 4.86) = 0.0430$$

$$\Sigma_{01} = \frac{1}{5-1} \sum_{k=0}^{4} (x_{k0} - 4.86)(x_{k1} - 3.28) = 0.0365$$

$$\Sigma_{02} = \frac{1}{5-1} \sum_{k=0}^{4} (x_{k0} - 4.86)(x_{k2} - 1.40) = -0.0025$$

$$\Sigma_{03} = \frac{1}{5-1} \sum_{k=0}^{4} (x_{k0} - 4.86)(x_{k3} - 0.20) = 0.0$$

We can repeat this calculation for all the rows of Σ to give the complete covariance matrix:

$$\Sigma = \begin{bmatrix} 0.0430 & 0.0365 & -0.0025 & 0.0000 \\ 0.0365 & 0.0670 & -0.0025 & 0.0000 \\ -0.0025 & -0.0025 & 0.0050 & 0.0000 \\ 0.0000 & 0.0000 & 0.0000 & 0.0000 \end{bmatrix}$$

The elements along the diagonal represent the variance of the features of **X**. Notice that for the fourth feature of **X**, all the variances and covariances are zero. This makes sense because all the values for this feature in **X** are the same; there is no variance.

We can calculate the covariance matrix for a set of observations in code by using np.cov:

```
>>> print(np.cov(X, rowvar=False))
[[ 0.043   0.0365 -0.0025  0.    ]
 [ 0.0365  0.067  -0.0025  0.    ]
 [-0.0025 -0.0025  0.005   0.    ]
 [ 0.      0.      0.      0.    ]]
```

Notice that the call to np.cov includes rowvar=False. By default, np.cov expects each row of its argument to be a variable and the columns to be the observations of that variable. This is the opposite of the usual way a set of observations is typically stored in a matrix for deep learning. Therefore, we use the rowvar keyword to tell NumPy that the rows, not the columns, are the observations.

I claimed above that the diagonal of the covariance matrix returns the variances of the features in **X**. NumPy has a function, np.std, to calculate the standard deviation, and squaring the output of this function should give us the variances of the features by themselves. For **X**, we get

```
>>> print(np.std(X, axis=0)**2)
[0.0344 0.0536 0.004  0.    ]
```

These variances don't look like the diagonal of the covariance matrix. The difference is due to the $n - 1$ in the denominator of the covariance equation, Equation 6.8. By default, np.std calculates what is known as a biased estimate of the sample variance. This means that instead of dividing by $n - 1$, it divides by n. To get np.std to calculate the unbiased estimator of the variance, we need to add the ddof=1 keyword,

```
>>> print(np.std(X, axis=0, ddof=1)**2)
[0.043 0.067 0.005 0.    ]
```

then we'll get the same values as along the diagonal of Σ.

Now that we know how to calculate the covariance matrix, let's use it in a distance metric.

Mahalanobis Distance

Above, we represented a dataset by a matrix where the rows of the dataset are observations and the columns are the values of variables that make up each observation. In machine learning terms, the rows are the feature vectors. As we saw above, we can calculate the mean of each feature across all the observations, and we can calculate the covariance matrix. With these values, we can define a distance metric called the *Mahalanobis distance*,

$$D_M = \sqrt{(x - \mu)^\top \Sigma^{-1}(x - \mu)} \qquad (6.9)$$

where x is a vector, μ is the vector formed by the mean values of each feature, and Σ is the covariance matrix. Notice that this metric uses the *inverse* of the covariance matrix, not the covariance matrix itself.

Equation 6.9 is, in some sense, measuring the distance between a vector and a distribution with the mean vector μ. The dispersion of the distribution is captured in Σ. If there is no covariance between the features in the dataset and each feature has the same standard deviation, then Σ becomes the identity matrix, which is its own inverse. In that case, Σ^{-1} effectively drops out of Equation 6.9, and the Mahalanobis distance becomes the L2-distance (Euclidean distance).

Another way to think of the Mahalanobis distance is to replace μ with another vector, call it y, that comes from the same dataset as x. Then D_M is the distance between the two vectors, taking the variance of the dataset into account.

We can use the Mahalanobis distance to build a simple classifier. If, given a dataset, we calculate the mean feature vector of each class in the dataset (this vector is also called the *centroid*), we can use the Mahalanobis distance to assign a label to an unknown feature vector, x. We can do so by calculating all the Mahalanobis distances to the class centroids and assigning x to the class returning the smallest value. This type of classifier is sometimes called a *nearest centroid* classifier, and you'll often see it implemented using the L2-distance in place of the Mahalanobis distance. Arguably, you can expect the Mahalanobis distance to be the better metric because it takes the variance of the dataset into account.

Let's use the breast cancer dataset included with sklearn to build the nearest centroid classifier using the Mahalanobis distance. The breast cancer dataset has two classes: benign (0) and malignant (1). The dataset contains 569 observations, each of which has 30 features derived from histology slides. We'll build two versions of the nearest centroid classifier: one using the Mahalanobis distance and the other using the Euclidean distance. Our expectation is that the classifier using the Mahalanobis distance will perform better.

The code we need is straightforward:

```
import numpy as np
from sklearn import datasets
❶ from scipy.spatial.distance import mahalanobis

bc = datasets.load_breast_cancer()
d = bc.data; l = bc.target
❷ i = np.argsort(np.random.random(len(d)))
d = d[i]; l = l[i]
xtrn, ytrn = d[:400], l[:400]
xtst, ytst = d[400:], l[400:]
```

```
❸ i = np.where(ytrn == 0)
   m0 = xtrn[i].mean(axis=0)
   i = np.where(ytrn == 1)
   m1 = xtrn[i].mean(axis=0)
   S = np.cov(xtrn, rowvar=False)
   SI= np.linalg.inv(S)

   def score(xtst, ytst, m, SI):
       nc = 0
       for i in range(len(ytst)):
           d = np.array([mahalanobis(xtst[i],m[0],SI),
                         mahalanobis(xtst[i],m[1],SI)])
           c = np.argmin(d)
           if (c == ytst[i]):
               nc += 1
       return nc / len(ytst)

   mscore = score(xtst, ytst, [m0,m1], SI)
❹ escore = score(xtst, ytst, [m0,m1], np.identity(30))
   print("Mahalanobis score = %0.4f" % mscore)
   print("Euclidean   score = %0.4f" % escore)
```

We start by importing the modules we need, including mahalanobis from
SciPy ❶. This function accepts two vectors and the inverse of a covariance
matrix and returns D_M. We get the dataset next in d with labels in 1. We randomize the order ❷ and pull out the first 400 observations as training data
(xtrn) with labels (ytrn). We hold back the remaining observations for testing
(xtst, ytst).

We *train* the model next. Training consists of pulling out all the observations belonging to each class ❸ and calculating m0 and m1. These are the
mean values of each of the 30 features for all class 0 and class 1 observations.
We then calculate the covariance matrix of the entire training set (S) and its
inverse (SI).

The score function takes the test observations, a list of the class mean
vectors, and the inverse of the covariance matrix. It runs through each test
observation and calculates the Mahalanobis distances (d). It then uses the
smallest distance to assign the class label (c). If the assigned label matches
the actual test label, we count it (nc). At the end of the function, we return
the overall accuracy.

We call the score function twice. The first call uses the inverse covariance matrix (SI), while the second call uses an identity matrix, thereby making score calculate the Euclidean distance instead. Finally, we print both
results.

The randomization of the dataset ❷ means that each time the code is
run, it will output slightly different scores. Running the code 100 times gives
the following mean scores ($\pm$ the standard deviation).

Distance	Mean score
Mahalanobis	0.9595 ± 0.0142
Euclidean	0.8914 ± 0.0185

This clearly shows that using the Mahalanobis distance leads to better model performance, with about a 7 percent improvement in accuracy.

One recent use of the Mahalanobis distance in deep learning is to take the top-level embedding layer values, a vector, and use the Mahalanobis distance to detect out-of-domain or adversarial inputs. An *out-of-domain input* is one that is significantly different from the type of data the model was trained to use. An *adversarial input* is one where an adversary is deliberately attempting to fool the model by supplying an input that isn't of class X but that the model will label as class X.

Kullback-Leibler Divergence

The *Kullback-Leibler divergence (KL-divergence)*, or *relative entropy*, is a measure of the similarity between two probability distributions: the lower the value, the more similar the distributions.

If P and Q are discrete probability distributions, the KL-divergence is

$$D_{KL}(P\|Q) = \sum_x P(x) \log_2 \left(\frac{P(x)}{Q(x)} \right)$$

where $\log_2$ is the logarithm base-2. This is an information-theoretic measure; the output is in bits of information. Sometimes the natural log, ln, is used, in which case the measure is said to be in *nats*. The SciPy function that implements the KL-divergence is in `scipy.special` as `rel_entr`. Note that `rel_entr` uses the natural log, not log base-2. Note also that the KL-divergence isn't a distance metric in the mathematical sense because it violates the symmetry property, $D_{KL}(P\|Q) \neq D_{KL}(Q\|P)$, but that doesn't stop people from using it as one from time to time.

Let's see an example of how we might use the KL-divergence to measure between different discrete probability distributions. We'll measure the divergence between two different binomial distributions and a uniform distribution. Then, we'll plot the distributions to see if, visually, we believe the numbers.

To generate the distributions, we'll take many draws from a uniform distribution with 12 possible outputs. We can do this quickly in code by using `np.random.randint`. Then, we'll take draws from two different binomial distributions, $B(12, 0.4)$ and $B(12, 0.9)$, meaning 12 trials with probabilities of 0.4 and 0.9 per trial. We'll generate histograms of the resulting draws, divide by the sum of the counts, and use the rescaled histograms as our probability distributions. We can then measure the divergences between them.

The code we need is

```
from scipy.special import rel_entr
N = 1000000
❶ p = np.random.randint(0,13,size=N)
❷ p = np.bincount(p)
❸ p = p / p.sum()
  q = np.random.binomial(12,0.9,size=N)
  q = np.bincount(q)
  q = q / q.sum()
  w = np.random.binomial(12,0.4,size=N)
  w = np.bincount(w)
  w = w / w.sum()
  print(rel_entr(q,p).sum())
  print(rel_entr(w,p).sum())
```

We load rel_entr from SciPy and set the number of draws for each distribution to 1,000,000 (N). The code to generate the respective probability distributions follows the same method for each distribution. We draw N samples from the distribution, starting with the uniform ❶. We use randint because it returns integers in the range $[0, 12]$ so we can match the discrete $[0, 12]$ values that binomial returns for 12 trials. We get the histogram from the vector of draws by using np.bincount. This function counts the frequency of unique values in a vector ❷. Finally, we change the counts into fractions by dividing the histogram by the sum ❸. This gives us a 12-element vector in p representing the probability that randint will return the values 0 through 12. Assuming randint uses a good pseudorandom number generator, we expect the probabilities to be roughly equal for each value in p. (NumPy uses the Mersenne Twister pseudorandom number generator, one of the better ones out there, so we're confident that we'll get good results.)

We repeat this process, substituting binomial for randint, sampling from binomial distributions using probabilities of 0.9 and 0.4. Again, histogramming the draws and converting the counts to fractions gives us the remaining probability distributions, q and w, based on 0.9 and 0.4, respectively.

We are finally ready to measure the divergence. The rel_entr function is a bit different from other functions in that it does not return D_{KL} directly. Instead, it returns a vector of the same length as its arguments, where each element of the vector is part of the overall sum leading to D_{KL}. Therefore, to get the actual divergence number, we need to add the elements of this vector. So, we print the sum of the output of rel_entr, comparing the two binomial distributions to the uniform distribution.

The random nature of the draws means we'll get slightly different numbers each time we run the code. One run gave

Distributions	Divergence
$D_{KL}(Q\|P)$	1.1826
$D_{KL}(W\|P)$	0.6218

This shows that the binomial distribution with probability 0.9 diverges more from a uniform distribution than the binomial distribution with probability 0.4. Recall, the smaller the divergence, the closer the two probability distributions are to each other.

Do we believe this result? One way to check is visually, by plotting the three distributions and seeing if $B(12, 0.4)$ looks more like a uniform distribution than $B(12, 0.9)$ does. This leads to Figure 6-1.

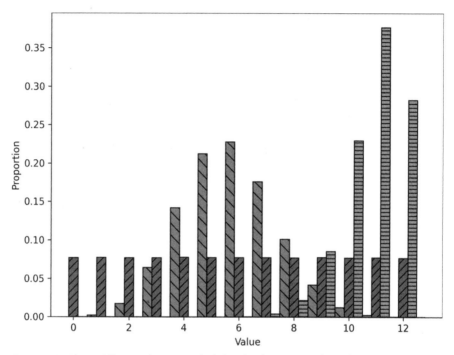

Figure 6-1: Three different, discrete probability distributions: uniform (forward hash), B(12,0.4) (backward hash), and B(12,0.9) (horizontal hash)

Although it is clear that neither binomial distribution is particularly uniform, the $B(12, 0.4)$ distribution is relatively centered in the range and spread across more values than the $B(12, 0.9)$ distribution is. It seems reasonable to think of $B(12, 0.4)$ as more like the uniform distribution, which is precisely what the KL-divergence told us by returning a smaller value.

We now have everything we need to implement principal component analysis.

Principal Component Analysis

Assume we have a matrix, X, representing a dataset. We understand that the variance of each of the features need not be the same. If we think of each observation as a point in an n-dimensional space, where n is the number of features in each observation, we can imagine a cloud of points with a different amount of scatter in different directions.

Principal component analysis (PCA) is a technique to learn the directions of the scatter in the dataset, starting with the direction aligned along the greatest scatter. This direction is called the *principal component*. You then find the remaining components in order of decreasing scatter, with each new component orthogonal to all the others. The top part of Figure 6-2 shows a 2D dataset and two arrows. Without knowing anything about the dataset, we can see that the largest arrow points along the direction of the greatest scatter. This is what we mean by the principal component.

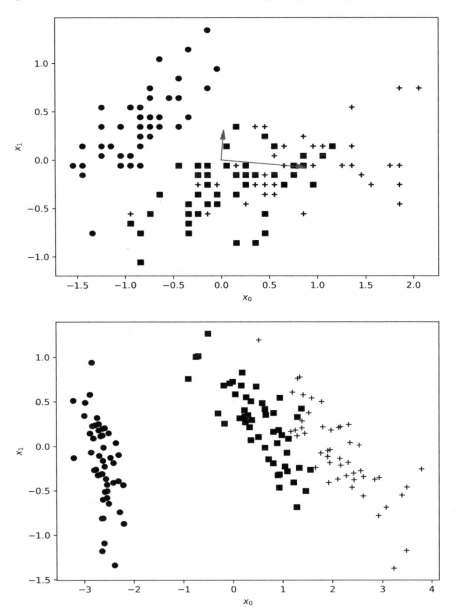

Figure 6-2: The first two features of the iris dataset and principal component directions (top), and the iris dataset after transformation by PCA (bottom)

We often use PCA to reduce the dimensionality of a dataset. If there are 100 variables per observation, but the first two principal components explain 95 percent of the scatter in the data, then mapping the dataset along those two components and discarding the remaining 98 components might adequately characterize the dataset with only two variables. We can use PCA to augment a dataset as well, assuming continuous features.

So, how does PCA work? All this talk about the scatter of the data implies that PCA might be able to make use of the covariance matrix, and, indeed, it does. We can break the PCA algorithm down into a few steps:

1. Mean center the data.

2. Calculate the covariance matrix, Σ, of the mean-centered data.

3. Calculate the eigenvalues and eigenvectors of Σ.

4. Sort the eigenvalues by decreasing absolute value.

5. Discard the weakest eigenvalues/eigenvectors (optional).

6. Form a transformation matrix, W, using the remaining eigenvectors.

7. Generate new transformed values from the existing dataset, $x' = Wx$. These are sometimes referred to as *derived variables*.

Let's work through an example of this process using the iris dataset (Listing 6-1). We'll reduce the dimensionality of the data from four features to two. First the code, then the explanation:

```
from sklearn.datasets import load_iris
iris = load_iris().data.copy()
❶ m = iris.mean(axis=0)
  ir = iris - m
❷ cv = np.cov(ir, rowvar=False)
❸ val, vec = np.linalg.eig(cv)
  val = np.abs(val)
❹ idx = np.argsort(val)[::-1]
  ex = val[idx] / val.sum()
  print("fraction explained: ", ex)
❺ w = np.vstack((vec[:,idx[0]],vec[:,idx[1]]))

❻ d = np.zeros((ir.shape[0],2))
  for i in range(ir.shape[0]):
      d[i,:] = np.dot(w,ir[i])
```

Listing 6-1: Principal component analysis (PCA)

We start by loading the iris dataset, courtesy of sklearn. This gives us iris as a 150 × 4 matrix, since there are 150 observations, each with four features. We calculate the mean value of each feature ❶ and subtract it from the dataset, relying on NumPy's broadcasting rules to subtract m from each row of iris. We'll work with the mean-centered matrix ir going forward.

The next step is to compute the covariance matrix ❷. The output, cv, is a 4 × 4 matrix, since we have four features per observation. We follow this by calculating the eigenvalues and eigenvectors of cv ❸ and then take the absolute value of the eigenvalues to get the magnitude. We want the eigenvalues in decreasing order of magnitude, so we get the indices that sort them that way ❹ using the Python idiom of [::-1] to reverse the order of a list or array.

The magnitude of the eigenvalues is proportional to the fraction of the variance in the dataset along each principal component; therefore, if we scale the eigenvalues by their overall sum, we get the proportion explained by each principal component (ex). The fraction of variance explained is

```
fraction explained: [0.92461872 0.05306648 0.01710261 0.00521218]
```

indicating that two principal components explain nearly 98 percent of the variance in the iris dataset. Therefore, we'll only keep the first two principal components going forward.

We create the transformation matrix, w, from the eigenvectors that go with the two largest eigenvalues ❺. Recall, eig returns the eigenvectors as the columns of the matrix vec. The transformation matrix, w, is a 2×4 matrix because it maps a four-component feature vector to a new two-component vector.

All that's left is to create a place to hold the transformed observations and fill them in ❻. The new, reduced-dimension dataset is in d. We can now plot the entire transformed dataset, labeling each point by the class to which it belongs. The result is the bottom part of Figure 6-2.

In the top part of Figure 6-2 is a plot of the original dataset using only the first two features. The arrows indicate the first two principal components, and the size of the arrows shows how much of the variance in the data these components explain. The first component explains most of the variance, which makes sense visually.

In this example, the derived variables in the bottom part of Figure 6-2 have made the dataset easier to work with, as the classes are better separated than on the top using only two of the original features. Sometimes, PCA makes it easier for a model to learn because of the reduced feature vector size. However, this is not always the case. During PCA, you may lose a critical feature allowing class separation. As with most things in machine learning, experimentation is vital.

PCA is commonly used and is therefore well supported in multiple toolkits. Instead of the dozen or so lines of code we used above, we can accomplish the same thing by using the PCA class from the sklearn.decomposition module:

```
from sklearn.decomposition import PCA
pca = PCA(n_components=2)
pca.fit(ir)
d = pca.fit_transform(ir)
```

The new, reduced-dimension dataset is in d. Like other sklearn classes, after we tell PCA how many components we want it to learn, it uses fit to set up the transformation matrix (w in Listing 6-1). We then apply the transform by calling fit_transform.

Singular Value Decomposition and Pseudoinverse

We'll end this chapter with an introduction to *singular value decomposition (SVD)*. This is a powerful technique to factor any matrix into the product of three matrices, each with special properties. The derivation of SVD is beyond the scope of this book. I trust motivated readers to dig into the vast literature on linear algebra to locate a satisfactory presentation of where SVD comes from and how it is best understood. Our goal is more modest: to become familiar with the mathematics found in deep learning. Therefore, we'll content ourselves with the definition of SVD, some idea of what it gives us, some of its uses, and how to work with it in Python. For deep learning, you'll most likely encounter SVD when calculating the pseudoinverse of a nonsquare matrix. We'll also see how that works in this section.

The output of SVD for an input matrix, A, with real elements and shape $m \times n$, where m does not necessarily equal n (though it could) is

$$A = U\Sigma V^{\top} \tag{6.10}$$

A has been decomposed into three matrices: U, Σ, and V. Note that you might sometimes see $V^{\top}$ written as V^{*}, the conjugate transpose of V. This is the more general form that works with complex-valued matrices. We'll restrict ourselves to real-valued matrices, so we only need the ordinary matrix transpose.

The SVD of an $m \times n$ matrix, A, returns the following: U, which is $m \times m$ and orthogonal; Σ, which is $m \times n$ and diagonal; and V, which is $n \times n$ and orthogonal. Recall that the transpose of an orthogonal matrix is its inverse, so $UU^{\top} = I_m$ and $VV^{\top} = I_n$, where the subscript on the identity matrix gives the order of the matrix, $m \times m$ or $n \times n$.

At this point in the chapter, you may have raised an eyebrow at the statement "Σ, which is $m \times n$ and diagonal," since we've only considered square matrices to be diagonal. Here, when we say *diagonal*, we mean a *rectangular diagonal matrix*. This is the natural extension to a diagonal matrix, where the elements of what would be the diagonal are nonzero and all others are zero. For example,

$$M = \begin{bmatrix} 1 & 0 & 0 & 0 & 0 \\ 0 & 2 & 0 & 0 & 0 \\ 0 & 0 & 3 & 0 & 0 \end{bmatrix}$$

is a 3×5 rectangular diagonal matrix because only the main diagonal is nonzero. The "singular" in "singular value decomposition" comes from the fact that the elements of the diagonal matrix, Σ, are the singular values, the square roots of the positive eigenvalues of the matrix $A^T A$.

SVD in Action

Let's be explicit and use SVD to decompose a matrix. Our test matrix is

$$A = \begin{bmatrix} 3 & 2 & 2 \\ 2 & 3 & -2 \end{bmatrix}$$

We'll show SVD in action as a series of steps. To get the SVD, we use svd from scipy.linalg,

```
>>> from scipy.linalg import svd
>>> a = np.array([[3,2,2],[2,3,-2]])
>>> u,s,vt = svd(a)
```

where u is U, vt is $V^\top$, and s contains the singular values:

```
>>> print(u)
[[-0.70710678 -0.70710678]
 [-0.70710678  0.70710678]]
>>> print(s)
[5. 3.]
>>> print(vt)
[[-7.07106781e-01 -7.07106781e-01 -5.55111512e-17]
 [-2.35702260e-01  2.35702260e-01 -9.42809042e-01]
 [-6.66666667e-01  6.66666667e-01  3.33333333e-01]]
```

Let's check that the singular values are indeed the square roots of the positive eigenvalues of $A^T A$:

```
>>> print(np.linalg.eig(a.T @ a)[0])
[2.5000000e+01 5.0324328e-15 9.0000000e+00]
```

This shows us that, yes, 5 and 3 are the square roots of 25 and 9. Recall that eig returns a list, the first element of which is a vector of the eigenvalues. Also note that there is a third eigenvalue: zero. You might ask: "How small a numeric value should we interpret as zero?" That's a good question with no hard and fast answer. Typically, I interpret values below 10^{-9} to be zero.

The claim of SVD is that U and V are unitary matrices. If so, their products with their transposes should be the identity matrix:

```
>>> print(u.T @ u)
[[1.00000000e+00 3.33066907e-16]
 [3.33066907e-16 1.00000000e+00]]
>>> print(vt @ vt.T)
[[ 1.00000000e+00  8.00919909e-17 -1.85037171e-17]
 [ 8.00919909e-17  1.00000000e+00 -5.55111512e-17]
 [-1.85037171e-17 -5.55111512e-17  1.00000000e+00]]
```

Given the comment above about numeric values that we should interpret as zero, this is indeed the identity matrix. Notice that svd returned $V^\top$, not V. However, since $(V^\top)^\top = V$, we're still multiplying $V^\top V$.

The svd function returns not Σ but the diagonal values of Σ. Therefore, let's reconstruct Σ and use it to see that SVD works, meaning we can use U, Σ, and $V^{\top}$ to recover A:

```
>>> S = np.zeros((2,3))
>>> S[0,0], S[1,1] = s
>>> print(S)
[[5. 0. 0.]
 [0. 3. 0.]]
>>> A = u @ S @ vt
>>> print(A)
[[ 3.  2.  2.]
 [ 2.  3. -2.]]
```

This is the A we started with—almost: the recovered A is no longer of integer type, a subtle change worth remembering when writing code.

Two Applications

SVD is a cute trick, but what can we do with it? The short answer is "a lot." Let's see two applications. The first is using SVD for PCA. The sklearn PCA class we used in the previous section uses SVD under the hood. The second example shows up in deep learning: using SVD to calculate the Moore-Penrose pseudoinverse, a generalization of the inverse of a square matrix to $m \times n$ matrices.

SVD for PCA

To see how to use SVD for PCA, let's use the iris data from the previous section so we can compare with those results. The key is to truncate the Σ and $V^{\top}$ matrices to keep only the desired number of largest singular values. The decomposition code will put the singular values in decreasing order along the diagonal of Σ for us, we need only retain the first k columns of Σ. In code, then,

```
  u,s,vt = svd(ir)
❶ S = np.zeros((ir.shape[0], ir.shape[1]))
  for i in range(4):
      S[i,i] = s[i]
❷ S = S[:, :2]
  T = u @ S
```

Here, we're using ir from Listing 6-1. This is the mean-centered version of the iris dataset matrix, with 150 rows of four features each. A call to svd gives us the decomposition of ir. The next three lines ❶ create the full Σ matrix in S. Because the iris dataset has four features, the s vector that svd returns will have four singular values.

The truncation comes by keeping the first two columns of S ❷. Doing this changes Σ from a 150×4 matrix to a 150×2 matrix. Multiplying U by the new Σ gives us the transformed iris dataset. Since U is 150×150 and Σ is 150×2, we get a 150×2 dataset in T. If we plot this as T[:,0] versus T[:,1], we get the exact same plot as the bottom part of Figure 6-2.

The Moore-Penrose Pseudoinverse

As promised, our second application is to compute A^+, the Moore-Penrose pseudoinverse of an $m \times n$ matrix A. The matrix A^+ is called a pseudo-inverse because, in conjunction with A, it acts like an inverse in that

$$AA^+A = A \tag{6.11}$$

where AA^+ is somewhat like the identity matrix, making A^+ somewhat like the inverse of A.

Knowing that the pseudoinverse of a rectangular diagonal matrix is simply the reciprocal of the diagonal values, leaving zeros as zero, followed by a transpose, we can calculate the pseudoinverse of any general matrix as

$$A^+ = V\Sigma^+U^*$$

for $A = U\Sigma V^*$, the SVD of A. Notice, we're using the conjugate transpose, V^*, instead of the ordinary transpose, $V^\top$. If A is real, then the ordinary transpose is the same as the conjugate transpose.

Let's see if the claim regarding A^+ is true. We'll start with the A matrix we used in the section above, compute the SVD, and use the parts to find the pseudoinverse. Finally, we'll validate Equation 6.11.

We'll start with A, the same array we used above for the SVD example:

```
>>> A = np.array([[3,2,2],[2,3,-2]])
>>> print(A)
[[ 3  2  2]
 [ 2  3 -2]]
```

Applying SVD will give us U and $V^\top$ along with the diagonal of Σ. We'll use the diagonal elements to construct Σ^+ by hand. Recall, Σ^+ is the transpose of Σ, where the diagonal elements that are not zero are changed to their reciprocals:

```
>>> u,s,vt = svd(A)
>>> Splus = np.array([[1/s[0],0],[0,1/s[1]],[0,0]])
>>> print(Splus)
[[0.2        0.        ]
 [0.         0.33333333]
 [0.         0.        ]]
```

Now we can calculate A^+ and verify that $AA^+A = A$:

```
>>> Aplus = vt.T @ Splus @ u.T
>>> print(Aplus)
```

```
[[ 0.15555556  0.04444444]
 [ 0.04444444  0.15555556]
 [ 0.22222222 -0.22222222]]
>>> print(A @ Aplus @ A)
[[ 3.  2.  2.]
 [ 2.  3. -2.]]
```

And, in this case, AA^+ is the identity matrix:

```
>>> print(A @ Aplus)
[[1.00000000e+00 5.55111512e-17]
 [1.66533454e-16 1.00000000e+00]]
```

This concludes our whirlwind look at SVD and our discussion of linear algebra. We barely scratched the surface, but we've covered what we need to know.

Summary

This heavy chapter and Chapter 5 before it plowed through a lot of linear algebra. As a mathematical topic, linear algebra is vastly richer than our presentation here.

We focused the chapter on square matrices, as they have a special place in linear algebra. Specifically, we discussed general properties of square matrices, with examples. We learned about eigenvalues and eigenvectors, how to find them, and why they are useful. Next, we looked at vector norms and other ways to measure distance, as they show up often in deep learning. Finally, we ended the chapter by learning what PCA is and how it works, followed by a look at singular value decomposition, with two applications relevant to deep learning.

The next chapter shifts gears and covers differential calculus. This is, fortunately, the "easy" part of calculus, and is, in general, all that we need to understand the algorithms specific to deep learning. So, fasten your seat belts, make sure your arms and legs are fully within the vehicle, and prepare for departure to the world of differential calculus.

7

DIFFERENTIAL CALCULUS

The discovery of "the calculus" by Sir Issac Newton, and separately by Gottfried Wilhelm Leibniz, was one of the greatest achievements in the history of mathematics. Calculus is typically split into two main parts: differential and integral. Differential calculus talks about rates of change and their relationships, embodied in the notion of the derivative. Integral calculus is concerned with things like the area under a curve.

We don't need integral calculus for deep learning, but we'll use differential calculus often. For example, we use differential calculus to train neural networks; we adjust the weights of a neural network using gradient descent, which relies on derivatives calculated via the backpropagation algorithm.

The derivative will be the star of this chapter. We'll begin by introducing the idea of slope and seeing how it leads to the notion of a derivative. We'll then formally define the derivative and learn how to calculate derivatives of functions of one variable. After that, we'll learn how to use derivatives to find the minima and maxima of functions. Next come partial derivatives, the derivatives with respect to a single variable for functions of more than one variable. We'll use partial derivatives extensively in the backpropagation algorithm. We'll conclude with gradients, which will introduce us to matrix calculus, the subject of Chapter 8.

Slope

In algebra class, we learned all about lines. One way to define a line is the slope-intercept form,

$$y = mx + b$$

where m is the slope and b is the y intercept, the place where the line crosses the y-axis. We're interested in the slope. If we know two points on the line, (x_1, y_1) and (x_0, y_0), we know the slope of the line:

$$m = \frac{y_1 - y_0}{x_1 - x_0}$$

The *slope* tells us how much of a change in y we will get for any given change in x position. If the slope is positive, then a positive change in x leads to a positive change in y. On the other hand, a negative slope means that a positive change in x leads to a negative change in y.

The slope-intercept form of the line tells us that the slope is the proportionality constant between x and y. The intercept, b, is a constant offset. This means a change in x position from x_1 to x_0 leads to an $m(x_1 - x_0)$ change in y. Slopes relate two things, telling us how changing one affects the other. We'll get back to this idea several times in the book.

Now, let's visualize this with some examples. Figure 7-1 shows the plot of a curve and some lines that intersect it.

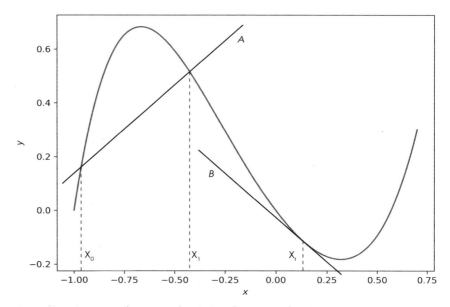

Figure 7-1: A curve with a secant line (A) and a tangent line (B)

The line labeled *A* crosses the curve at two points, x_1 and x_0. A line passing between two points on a curve is called a *secant* line. The other line, *B*, just touches the curve at the point x_t. Lines that touch a curve at one point are called *tangent* lines. We'll get back to secant lines in the next section, but, for now, notice that the tangent line has a particular slope at x_t and that a secant line becomes a tangent line as the distance between the points x_1 and x_0 goes toward zero.

Imagine that we move the point x_t from place to place along the curve; we can see that the slope of the tangent line at x_t would change with it. As we approach the minimum point of the curve, near $x = 0.3$, we see that the slope becomes more and more shallow. If we approach from the left, the slope is negative and becomes less and less negative. If we approach from the right, the slope is positive but becomes smaller and smaller. At the actual minimum point, near $x = 0.3$, the tangent line is horizontal, with a slope of zero. Similarly, if we approach the maximum of the curve, near $x = -0.8$, the slope also approaches zero.

We can see that the tangent line tells us how the curve is changing at a point. As we'll see later in the chapter, the fact that the slope of a tangent line is zero at the minima and maxima of a curve points us toward a method for finding these points. The points where the slope of the tangent line is zero are known as *stationary points*.

Of course, to take advantage of the slope of the tangent line, we need to be able to find its value for any x on the curve. The next section will show us how.

Derivatives

The previous section introduced the idea of secant and tangent lines and hinted that knowing the slope of the tangent line at any point on a curve is a potentially useful thing. The slope of the tangent line at a point x is known as the *derivative* at x. It tells us how the curve (function) is changing at the point x, that is, how the function value changes with an infinitesimal change in x. In this section, we'll formally define the derivative and learn shortcut rules for calculating derivatives of functions of a single variable, x.

A Formal Definition

A typical first-semester calculus course introduces you to derivatives through studying limits. I mentioned above how the slope of the secant line between two points on a curve becomes a tangent line when the points collapse on top of each other, and that's one place where limits come into play.

For example, if $y = f(x)$ is a curve, and we have two points on the curve, x_0 and x_1, then the slope, $\Delta y / \Delta x$, between these points is

$$\frac{\Delta y}{\Delta x} = \frac{y_1 - y_0}{x_1 - x_0} = \frac{f(x_1) - f(x_0)}{x_1 - x_0} \tag{7.1}$$

This is the *rise over run* you may remember learning in school. The *rise*, $\Delta y = y_1 - y_0 = f(x_1) - f(x_0)$, is divided by the *run*, $\Delta x = x_1 - x_0$. We typically use Δ as a prefix to mean the change in some variable.

If we define $h = x_1 - x_0$, we can rewrite Equation 7.1 as

$$\frac{\Delta y}{\Delta x} = \frac{f(x_0 + h) - f(x_0)}{h}$$

since $x_1 = x_0 + h$.

In this new form, we can find the slope of the tangent line at x_0 by letting h get closer and closer to zero, $h \to 0$. Letting a value approach another value is a *limit*. Letting $h \to 0$ moves the two points we're calculating the slope between closer and closer. This leads directly to the definition of the derivative

$$\frac{dy}{dx} = f'(x) \equiv \lim_{h \to 0} \frac{f(x + h) - f(x)}{h} \tag{7.2}$$

where dy/dx or $f'(x)$ is used to represent the derivative of $f(x)$.

Before we get into what the derivative means, let's take a minute to discuss notation. Using $f'(x)$ for the derivative follows Joseph-Louis Lagrange. Leibniz used dy/dx to mirror the notation for slope with $\Delta \to d$. If Δy is a change in y between two points, dy is the infinitesimal change in y at a single point. Newton used yet another notation, $\dot{f}$, with the dot representing the derivative of f. Physicists often use Newton's notation for the specific case of derivatives with respect to time. For example, if $f(t)$ is the position of a particle as a function of time, t, then $\dot{f}(t)$ is the derivative with respect to t, that is, how the position is changing in time. How a position changes in time is the speed (velocity if using vectors). You'll see all of these notations in books. My preference is to preserve $\dot{f}$ for functions of time and use Lagrange's $f'(x)$ and Leibniz's dy/dx interchangeably elsewhere.

Although Equation 7.2 above is quite tedious and the bane of many beginning calculus students, at least until they hit integration, you could work with it if you had to. However, we won't discuss integration at all in this book, so you can take a deep breath and relax.

After the struggle with limits, calculus students are let in on a secret: a small set of rules will allow you to calculate virtually all derivatives *without* using limits. We'll start by introducing these rules, one at a time with examples, and then, at the end of this section, we'll put them all together in a form suitable for a T-shirt.

However, before we dive into the rules, let's spend a little more time discussing *what* the derivative is telling us. Above, I mentioned that how a position changes in time is given by the derivative. This is true of all derivatives; they tell us how something is changing with respect to how something else is changing. We even see this in Leibniz's notation, dy/dx, how dy changes for a change in dx. The derivative at x tells us how the function is changing at x. As we'll see, the derivative of $f(x)$ is itself a new function of x. If we pick a specific x_0, then we know that $f(x_0)$ is the value of the function at x_0.

Similarly, if we know the derivative, then $f'(x_0)$ is how quickly, and in which direction, the function, $f(x)$, is changing at x_0. Consider the definition of speed as how the position changes with time. We even say it in words: my current speed is 30 mph—*miles per hour*—a change in position with respect to a change in time.

We'll use derivatives like rates and see how changing one thing affects another. In the end, for deep learning, we want to know how changing the value of a parameter in a network will ultimately change the loss function, the error between what the network should have output and what it actually output.

If $f'(x)$ is a function of x, then we should be able to take the derivative of it. We call $f'(x)$ the *first derivative*. Its derivative, which we denote as $f''(x)$, is the *second derivative*. In Leibniz's notation, we write d^2y/dx^2. The second derivative tells us how the first derivative is changing with respect to x. Physics helps here. The first derivative of the position as a function of time (f) is the *velocity*, $\dot{f}$—how the position is changing with time. Therefore, the second derivative of the position, $\ddot{f}$, which is the first derivative of the velocity, is how the velocity is changing with time. We call this the *acceleration*.

In theory, there is no end to how many derivatives we can calculate. In reality, many functions ultimately end up with a derivative that is a constant value. Since a constant value doesn't change, its derivative is zero.

To sum up, the derivative of $f(x)$ is another function, $f'(x)$ or dy/dx, that tells us the slope of the line tangent to $f(x)$ at every point. And, since $f'(x)$ is a function of x, it also has a derivative, $f''(x)$ or d^2y/dx^2, the second derivative, telling us how $f'(x)$ changes at each x, and so on. We'll see below how to make use of first and second derivatives. For now, let's learn the rules of *differentiation*, the act of calculating a derivative.

Basic Rules

We mentioned one rule in the previous section: that the derivative of a constant, c, is zero. So, we write

$$\frac{d}{dx}c = 0$$

where we're using Leibniz's notation in operator form: d/dx. Think of d/dx as something operating on what follows; it does this the same way that negation does: to negate c, we write $-c$; to differentiate c, we write $\frac{d}{dx}c$. If our expression has no x, then we will treat it as a constant, and the derivative will be zero.

The Power Rule

The derivative of a power of x uses the *power rule*,

$$\frac{d}{dx}ax^n = nax^{n-1}$$

where a is a constant and n is an exponent that doesn't need to be an integer. Let's see some examples:

$$\frac{d}{dx}x^3 = 3x^2$$

$$\frac{d}{dx}4x^2 = (2)4x^{2-1} = 8x$$

$$\frac{d}{dx}x = (1)x^{1-1} = (1)x^0 = 1$$

$$\frac{d}{dx}\sqrt{x} = \frac{d}{dx}x^{\frac{1}{2}} = \frac{1}{2}x^{-\frac{1}{2}} = \frac{1}{2\sqrt{x}}$$

$$\frac{d}{dx}0.1x^{0.07} = 0.007x^{-0.93} = \frac{0.007}{x^{0.93}}$$

We often build algebraic expressions out of terms that are added and subtracted. Differentiation is a linear operator, so we can write

$$\frac{d}{dx}(f(x) \pm g(x)) = \frac{d}{dx}f(x) \pm \frac{d}{dx}g(x)$$

This means we calculate derivatives term by term. For example, with the set of rules we have so far, we now know how to calculate the derivative of a polynomial:

$$\frac{d}{dx}(3x^4 - 2x^2 + 3x + 4) = 12x^3 - 4x + 3$$

$$\frac{d}{dx}(x^5 - 7x^2 + 42) = 5x^4 - 14x$$

In general, then,

$$\frac{d}{dx}(ax^n + bx^{n-1} + cx^{n-2} + \ldots + yx + z) = anx^{n-1} + b(n-1)x^{n-2} +$$

$$c(n-2)x^{n-3} + \ldots + y$$

where we see that the derivative of a polynomial of degree n is another polynomial of degree $n-1$ and that any constant term in the original polynomial drops to zero.

The Product Rule

Differentiation of functions multiplied together has its own rule, the *product rule*:

$$\frac{d}{dx}f(x)g(x) = f'(x)g(x) + f(x)g'(x)$$

The derivative of the product is the derivative of the first function times the second plus the derivative of the second times the first. Consider the following examples:

$$\frac{d}{dx}(x^2 - 4)(3x + 3) = (2x)(3x + 3) + (x^2 - 4)(3)$$

$$= 9x^2 + 6x - 12$$

$$\frac{d}{dx}(x - 3)(4x + 5) = (1)(4x + 5) + (x - 3)(4)$$

$$= 8x - 7$$

$$\frac{d}{dx}(2x + 2)(x^2 - x - 3) = (2)(x^2 - x - 3) + (2x + 2)(2x - 1)$$

$$= 6x^2 - 8$$

The Quotient Rule

The derivative of a function divided by another function follows the *quotient rule*:

$$\frac{d}{dx}\left(\frac{f(x)}{g(x)}\right) = \frac{f'(x)g(x) - f(x)g'(x)}{[g(x)]^2}$$

This leads to examples like these:

$$\frac{d}{dx}\left(\frac{5x-3}{2x+1}\right) = \frac{5(2x+1)-(5x-3)(2)}{(2x+1)^2}$$

$$= \frac{11}{(2x+1)^2}$$

$$\frac{d}{dx}\left(\frac{x^2+2x-3}{x^3-9}\right) = \frac{(2x+2)(x^3-9)-(x^2+2x-3)(3x^2)}{(x^3-9)^2}$$

$$= \frac{x^4+4x^3-9x^2+18x+18}{(x^3-9)^2}$$

The Chain Rule

The next rule concerns the composition of functions. Two functions are composed when the output of one is used as the input to another. The *chain rule* applies to function compositions, and it's of fundamental importance in the training of neural networks. The rule is

$$\frac{d}{dx}f(g(x)) = f'(g(x))g'(x)$$

We multiply the derivative of the outer function, using $g(x)$ as the variable, by the derivative of the inner function with respect to x.

As a first example, consider the function $f(x) = (x^2 + 2x + 3)^2$. Is this the composition of two functions? It is. Let's define $f(g) = g^2$ and $g(x) = x^2 + 2x + 3$. Then, we can find $f(x)$ by replacing every instance of g in $f(g)$ with the definition of g in terms of x: $g(x) = x^2 + 2x + 3$. This gives

$$f(x) = g^2 = (x^2 + 2x + 3)^2$$

which is, naturally, what we started with. To find $f'(x)$, we first find $f'(g)$, the derivative of f with respect to g, and then multiply by $g'(x)$, the derivative of g with respect to x. As a final step, we replace references to g with its definition in terms of x. So, we calculate $f'(x)$ as

$$f'(x) = (2g)(2x+2)$$

$$= 2(x^2+2x+3)(2x+2)$$

$$= 4(x+1)(x^2+2x+3)$$

We typically don't explicitly call out $f(g)$ and $g(x)$, but mentally we work through the same process. Let's see some more examples. In this one, we want to find

$$\frac{d}{dx}(2(4x-5)^2+3)$$

If we use the chain rule, we see that we have $f(g) = 2g^2 + 3$ and $g(x) = 4x - 5$. Therefore, we can write

$$\frac{d}{dx}(2(4x-5)^2+3) = (4g)(4)$$

$$= 4(4x-5)(4)$$

$$= 16(4x-5)$$

where $f'(g) = 2g$ and $g'(x) = 4$. With some practice, we'd mentally picture the $4x - 5$ of $f(x)$ as its own variable (the substitution of g) and then remember to multiply the derivative of $4x - 5$ when done. What if we didn't see the composition? What if we expanded the entire function, $f(x)$, and then took the derivative? We'd better get the answer we found above by using the chain rule. Let's see . . .

$$f(x) = 2(4x-5)^2+3$$

$$= 2(16x^2 - 40x + 25) + 3$$

$$= 32x^2 - 80x + 53$$

$$f'(x) = 64x - 80$$

$$= 16(4x-5)$$

This is what we found above, so our application of the chain rule is correct.

Let's look at another example. If $f(x) = \frac{1}{3x^2}$, how should we think of calculating the derivative? If we look at the function as $f(x) = \frac{u(x)}{v(x)}$, with $u(x) = 1$ and $v(x) = 3x^2$, we can use the quotient rule and get the following.

$$f'(x) = \frac{u'v - uv'}{v^2}$$

$$= \frac{0(3x^2) - 1(6x)}{(3x^2)^2}$$

$$= \frac{-6x}{9x^4}$$

$$= -\frac{2}{3x^3}$$

Here, we used a shorthand notation with u and v that drops the formal function of x notation.

We can also picture $f(x)$ as $(3x^2)^{-1}$. If we think of it this way, we can apply the chain rule and power rule to get

$$f'(x) = (-1)(3x^2)^{-2}(6x)$$

$$= \frac{-6x}{9x^4}$$

$$= -\frac{2}{3x^3}$$

proving that sometimes there's more than one way to calculate a derivative.

We presented the chain rule using Lagrange's notation. Later in the chapter, we'll see it again using Leibniz's notation. Let's move on now and present a set of rules for trigonometric functions.

Rules for Trigonometric Functions

The derivatives of the basic trigonometric functions are straightforward:

$$\frac{d}{dx} \sin x = \cos x$$

$$\frac{d}{dx} \cos x = -\sin x$$

$$\frac{d}{dx} \tan x = \sec^2 x$$

We can see the last rule is correct if we apply the basic differentiation rules to the definition of the tangent:

$$\frac{d}{dx}\tan x = \frac{d}{dx}\left(\frac{\sin x}{\cos x}\right)$$

$$= \frac{\cos x \cos x - \sin x(-\sin x)}{\cos^2 x}$$

$$= \frac{\sin^2 x + \cos^2 x}{\cos^2 x}$$

$$= \frac{1}{\cos^2 x}$$

$$= \sec^2 x$$

Remember that $\sec x = 1/\cos x$ and $\sin^2 x + \cos^2 x = 1$.

Let's look at some examples using the new trig rules. We'll start with one composing a function with a trig function:

$$\frac{d}{dx}\sin(x^3 - 3x) = \cos(x^3 - 3x)(3x^2 - 3)$$

$$= 3(x^2 - 1)\cos(x^3 - 3x)$$

We can see that this is a composition of $f(g) = \sin(g)$ and $g(x) = x^3 - 3x$, so we know we can apply the chain rule to get the derivative as $f'(g)g'(x)$ with $f'(g) = \cos(g)$ and $g'(x) = 3x^2 - 3$. The second line simplifies the answer.

Let's look at a more complicated composition:

$$\frac{d}{dx}\sin^2(x^3 - 3x) = 2\sin(x^3 - 3x)\cos(x^3 - 3x)(3x^2 - 3)$$

$$= 6(x^2 - 1)\sin(x^3 - 3x)\cos(x^3 - 3x)$$

This time, we break up the composition as $f(g) = g^2$ and $g(x) = \sin(x^3 - 3x)$. However, $g(x)$ is itself a composition of $g(u) = \sin(u)$ and $u(x) = x^3 - 3x$, as we had in the previous example. So, the first step is to write the following.

$$f'(x) = f'(g(x))g'(x)$$

$$= 2g(x)g'(x)$$

$$= 2\sin(x^3 - 3x)g'(x)$$

The first line is the definition of the derivative of a composition. The second line substitutes the derivative of $f(g)$, which is $2g$, and the third line replaces $g(x)$ with $\sin(x^3 - 3x)$. Now, we just need to find $g'(x)$, which we can do by using the chain rule a second time with $g(u) = \sin u$ and $u(x) = x^3 - 3x$, as we did for the example above. Doing this gives us $g'(x) = \cos(x^3 - 3x)(3x^2 - 3)$, so now we know that $f'(x)$ is

$$f'(x) = 2\sin(x^3 - 3x)g'(x)$$

$$= 2\sin(x^3 - 3x)\cos(x^3 - 3x)(3x^2 - 3)$$

$$= 6(x^2 - 1)\sin(x^3 - 3x)\cos(x^3 - 3x)$$

Let's do one more example. This one will involve more than one trig function. We want to see how to calculate

$$\frac{d}{dx}\left(\frac{\sin^3(x)}{\cos(x)}\right) = 2\sin^2(x) + \tan^2(x)$$

As is often the case when working with trig functions, identities come into play. Here, we see that $f(x)$ can be rewritten:

$$f(x) = \frac{\sin^3(x)}{\cos(x)}$$

$$= \sin^2(x)\left(\frac{\sin(x)}{\cos(x)}\right)$$

$$= \sin^2(x)\tan(x)$$

Now the derivative uses the trig rules, the definition of the secant, the chain rule, and the product rule, as shown next.

$$f'(x) = \frac{d}{dx}(\sin^2(x))\tan(x) + \sin^2(x)\frac{d}{dx}(\tan(x))$$

$$= 2\sin(x)\cos(x)\tan(x) + \sin^2(x)\sec^2(x)$$

$$= 2\sin(x)\cos(x)\tan(x) + \sin^2\left(\frac{1}{\cos^2(x)}\right)$$

$$= 2\sin(x)\cos(x)\left(\frac{\sin(x)}{\cos(x)}\right) + \tan^2(x)$$

$$= 2\sin^2(x) + \tan^2(x)$$

Let's move on and look at derivatives of exponentials and logarithms.

Rules for Exponentials and Logarithms

The derivative of e^x, where e is the base of the natural logarithm ($e \approx 2.718\dots$), is particularly simple. It is itself:

$$\frac{d}{dx}e^x = e^x$$

When the argument is a function of x, this becomes

$$\frac{d}{dx}e^{g(x)} = g'(x)e^{g(x)} \tag{7.3}$$

If the derivative of e^x is e^x, what is the derivative of a^x when a is a real number other than e? To see the answer, we need to remember that e^x and $\ln x$, the natural logarithm using base e, are inverse functions, so $e^{\ln a} = a$. Then, we can write

$$a^x = (e^{\ln a})^x = e^{x\ln a}$$

We know how to find the derivative of $e^{x\ln a}$ from Equation 7.3 above. It is

$$\frac{d}{dx}e^{x\ln a} = \ln(a)e^{x\ln a}$$

but $e^{x\ln a} = a^x$, so we have

$$\frac{d}{dx}a^x = a^x\ln a$$

and, in general,

$$\frac{d}{dx}a^{g(x)} = \ln(a)\,g'(x)a^{g(x)} \tag{7.4}$$

Notice, if $a = e$, we have $\ln(e) = 1$, and Equation 7.4 becomes Equation 7.3.

Let's look now at the derivative of the natural log itself. It is

$$\frac{d}{dx} \ln x = \frac{1}{x}$$

When the argument is a function of x, this becomes

$$\frac{d}{dx} \ln g(x) = \frac{g'(x)}{g(x)} \tag{7.5}$$

You may be wondering: How do we find the derivative of a logarithm that uses a base other than e? For example, what is the derivative of $\log_{10} x$? To answer this question, we do something similar to what we did above for the derivative of a^x. We write the logarithm of x for some base, b, in terms of the natural log, as

$$\log_b x = \frac{\ln x}{\ln b}$$

Here, $\ln b$ is a constant that does not depend on x. Also, we now know how to find the derivative of $\ln x$, so we see that the derivative of $\log_b x$ must be

$$\frac{d}{dx} \log_b x = \frac{1}{x \ln b}$$

for any real number base $b \neq 1$. And, still more generally,

$$\frac{d}{dx} \log_b g(x) = \frac{g'(x)}{g(x) \ln b} \tag{7.6}$$

where, again, we notice that if $b = e$, we get $\ln e = 1$, and Equation 7.6 becomes Equation 7.5.

With Equation 7.6, we've reached the end of our rules for derivatives. Let's now put them together in a single table we can refer back to throughout the remainder of the book. The result is Table 7-1.

We know how to find derivatives now. I would encourage you to look for practice sheets with worked-out answers to convince yourself that you understand the rules and how to apply them. Let's move on and look at how we can use derivatives to find the minima and maxima of functions. Finding minima is critical to the training of neural networks.

Table 7-1: The Rules of Differentiation

Type	Rule
Constants	$\dfrac{d}{dx}c = 0$
Powers	$\dfrac{d}{dx}ax^n = anx^{n-1}$
Sums	$\dfrac{d}{dx}f(x) \pm g(x) = f'(x) \pm g'(x)$
Products	$\dfrac{d}{dx}f(x)g(x) = f'(x)g(x) + f(x)g'(x)$
Quotients	$\dfrac{d}{dx}\left(\dfrac{f(x)}{g(x)}\right) = \dfrac{f'(x)g(x) - f(x)g'(x)}{[g(x)]^2}$
Chain	$\dfrac{d}{dx}f(g(x)) = f'(g(x))g'(x)$
Trigonometry	$\dfrac{d}{dx}\sin(g(x)) = g'(x)\cos(g(x))$
	$\dfrac{d}{dx}\cos(g(x)) = -g'(x)\sin(g(x))$
	$\dfrac{d}{dx}\tan(g(x)) = g'(x)\sec^2(g(x))$
Exponents	$\dfrac{d}{dx}e^{g(x)} = g'(x)e^{g(x)}$
	$\dfrac{d}{dx}a^{g(x)} = \ln(a)g'(x)a^{g(x)}$
Logarithms	$\dfrac{d}{dx}\ln g(x) = \dfrac{g'(x)}{g(x)}$
	$\dfrac{d}{dx}\log_b g(x) = \dfrac{g'(x)}{g(x)\ln b}$

Minima and Maxima of Functions

Earlier, I defined stationary points as places where the first derivative of a function is zero, that is, places where the slope of the tangent line is zero. We can use this information to decide if a particular point, call it x_m, is a minimum or maximum of the function, $f(x)$. If x_m is a minimum, it is a low point of the function, where $f(x_m)$ is smaller than any point to the immediate left or right of $f(x_m)$. Similarly, if $f(x_m)$ is higher than any point to the immediate left or right, $f(x_m)$ is a maximum. We collectively refer to minima and maxima as *extrema* of $f(x)$ (singular, *extremum*).

In terms of the derivative, a *minimum* is a place where the derivative of points immediately to the left of x_m are negative, and derivatives of points directly to the right of x_m are positive. *Maxima* are the reverse: derivatives to the left are positive, and derivatives to the right of x_m are negative.

Look back at Figure 7-1. There, we have a local maximum at about $x = -0.8$ and a local minimum at about $x = 0.3$. Let's say that the maximum is actually at $x_m = -0.8$. This is a maximum because if we look at any point x_p in the vicinity of x_m, $f(x_p)$ is less than $f(x_m)$. Likewise, if the minimum is at $x_m = 0.3$, that's because any point x_p near it has $f(x_p) > f(x_m)$. If we imagine the tangent line sliding along the graph, as it approaches $x = -0.8$, we see that the slope is positive but heading toward zero. If we move past $x = -0.8$, the slope is now negative. The reverse is true for the minimum at $x = 0.3$. Tangent lines approaching from the left have negative slope, but once they're past $x = 0.3$, the slope is positive.

You'll read and hear the terms *global* and *local* applied to minima and maxima. The global minimum of $f(x)$ is the lowest of all the minima of $f(x)$, and the global maximum is the highest of all the maxima. Other minima and maxima, then, are considered local; they are effective over a particular region, but there are other minima that are lower or maxima that are higher. We should note that not all functions have minima or maxima. For example, a line, $f(x) = mx + b$, has no minima or maxima, because there are no points on the line that satisfy the requirements for either.

So, if the first derivative, $f'(x)$, is zero, we have a minimum or maximum, right? Not so fast. At other stationary points, the first derivative may be zero, but the remaining criteria for a minimum or maximum are not met. These points are often called *inflection points* or, if in multiple dimensions, *saddle points*. For example, consider $y = x^3$. The first derivative is $y' = 3x^2$, and the second derivative is $y'' = 6x$. Both the first and second derivative are zero at $x = 0$. However, as we can see in Figure 7-2, the slope is positive to both the immediate left and immediate right of $x = 0$. Therefore, the slope never switches from positive to negative or negative to positive, meaning $x = 0$ is not an extremum but is an inflection point.

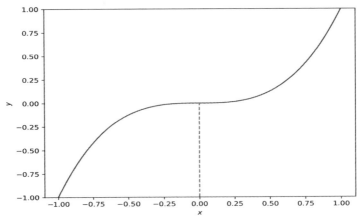

Figure 7-2: A graph of $y = x^3$ showing an inflection point at $x = 0$

Now assume that x_s is a stationary point so that $f'(x_s) = 0$. If we pick two other points, $x_{s-\epsilon}$ and $x_{s+\epsilon}$, one just to the left of x_s and the other just to the right, for some very small ϵ (epsilon), we have four possibilities for the values of $f'(x_{s-\epsilon})$ and $f'(x_{s+\epsilon})$, shown in Table 7-2.

Table 7-2: Identifying Stationary Points

Sign of $f'(x_s - \epsilon)$, $f'(x_s + \epsilon)$	Type of stationary point at $x_s - \epsilon < x_s < x_s + \epsilon$
+, –	Maximum
–, +	Minimum
+, +	Neither
–, –	Neither

Therefore, the value of the first derivative at a candidate stationary point isn't enough to tell us whether the point represents a minimum or maximum. We can look at the region around the candidate point to help us decide. We can also look at the value of $f''(x)$, the second derivative of $f(x)$. If x_s is a stationary point where $f'(x_s) = 0$, the sign of $f''(x_s)$ can tell us about what type of stationary point x_s might be. If $f''(x_s) < 0$, then x_s is a *maximum* of $f(x)$. If $f''(x_s) > 0$, x_s is a minimum. If $f''(x_s) = 0$, the second derivative isn't helpful; we will need to explicitly test nearby points with the first derivative.

How do we find candidate stationary points in the first place? For algebraic functions, we solve $f'(x) = 0$; we find the solution set of all the x values that make the first derivative of $f(x)$ zero. We then use the derivative tests to decide which are minima, maxima, or inflection points.

For many functions, we can find the solutions to $f'(x) = 0$ directly. For example, if $f(x) = x^3 - 2x + 4$, we have $f'(x) = 3x^2 - 2$. If we set this equal to zero, $3x^2 - 2 = 0$, and solve using the quadratic formula, we find that there are two stationary points: $x_0 = -\sqrt{6}/3$ and $x_1 = \sqrt{6}/3$. The second derivative of $f(x)$ is $f''(x) = 6x$. The sign of $f''(x_0)$ is negative; therefore, x_0 represents a maximum. And because the sign of $f''(x_1)$ is positive, x_1 is a minimum.

We can see that the derivative tests are correct. The top of Figure 7-3 shows us a plot of $f(x) = x^3 - 2x + 4$, where x_0 is a maximum and x_1 is a minimum.

Let's look at one more example. This time, we have $f(x) = x^5 - 2x^3 + x + 2$, the bottom plot of Figure 7-3. We find the first derivative and set it to zero:

$$f'(x) = 5x^4 - 6x^2 + 1 = 0$$

If we substitute $u = x^2$, we can solve for the roots of $f'(x)$ by finding the roots of $5u^2 - 6u + 1$ and setting those equal to x^2. Doing this gives us $u = 1, \frac{1}{5}$ and $x = \pm 1, \frac{\pm 1}{\sqrt{5}}$, so we have four stationary points. To test them, we can use the second derivative test. The second derivative is $f''(x) = 20x^3 - 12x$.

Substituting the stationary points into f'' gives

$$f''(x_0 = -1) = -8$$

$$f''(x_1 = \frac{-1}{\sqrt{5}}) \approx 3.5777$$

$$f''(x_2 = \frac{1}{\sqrt{5}}) \approx -3.5777$$

$$f''(x_3 = 1) = 8$$

meaning x_0 is a maximum, x_1 is a minimum, x_2 is another maximum, and x_3 is a minimum. Figure 7-3 again confirms our conclusions.

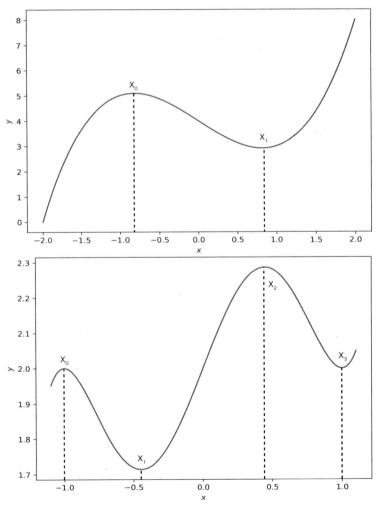

Figure 7-3: Plots of $f(x) = x^3 - 2x + 4$ (top) and $x^5 - 2x^3 + x + 2$ (bottom), with extrema marked

What if we can't easily find the stationary points of a function? Perhaps we can't solve the function algebraically, or maybe it can't be expressed in closed form, meaning no finite set of operations represents it. A typical calculus course isn't interested in these situations. Still, we need to be, because one way to think of a neural network is as a function approximator, one whose function can't be expressed directly. Can we still profitably use our new knowledge of derivatives? The answer is yes, we can. We can use the derivative as a pointer to tell us how to move closer and closer to the extrema. This is what gradient descent does, and we'll spend quite a bit of time discussing it later in the book.

For now, however, let's move on and examine functions of more than one variable and see what this does to the idea of a derivative.

Partial Derivatives

So far, we've focused exclusively on functions of one variable, x. What happens to the notion of differentiation when we have functions of more than one variable, say, $f(x, y)$, or $f(x_0, x_1, x_2, \ldots, x_n)$? To handle these cases, we'll introduce the idea of a *partial derivative*. Note that, for clarity, we'll use Leibniz's notation in this section.

Equation 7.2 defined the derivative of $f(x)$ with respect to x. If x is the only variable, why did we add the extra phrase "with respect to x"? Now we'll find out why: the partial derivative with respect to one of the variables in the expression is found by holding all the other variables fixed. We treat them as if they were constants. Then we say we're calculating the partial derivative with respect to the one not held fixed.

Let's look at an example. Let $f(x, y) = xy + x/y$. Then, we can calculate *two* partial derivatives, one with respect to x and the other with respect to y:

$$\frac{\partial f}{\partial x} = y + \frac{1}{y}$$

$$\frac{\partial f}{\partial y} = x + \frac{\partial}{\partial y} xy^{-1} = x - \frac{x}{y^2}$$

The rules of differentiation we learned earlier in the chapter still apply. Notice the d has changed to ∂. This indicates that the function, f, is of more than one variable. Also, see that when calculating the respective derivatives, we held the other variable fixed as if it were a parameter. As far as calculating partial derivatives, that's all there is to it. Let's see a few more examples to help you make the idea more concrete.

If $f(x, y, z) = x^2 + y^2 + z^2 + 3xyz$, we can find three partial derivatives:

$$\frac{\partial f}{\partial x} = 2x + 3yz$$

$$\frac{\partial f}{\partial y} = 2y + 3xz$$

$$\frac{\partial f}{\partial z} = 2z + 3xy$$

The other two variables are considered constant. This is why, for example, in the partial derivative with respect to x, y^2 and z^2 become 0, and $3xyz$ becomes $3yz$.

If $f(x, y, z, t) = \frac{xy}{zt} + y\sqrt{z} + \sqrt{xt}$, we have four partial derivatives:

$$\frac{\partial f}{\partial x} = \frac{y}{zt} + 0 + \frac{\partial}{\partial x}\sqrt{t}x^{\frac{1}{2}}$$

$$= \frac{y}{zt} + \frac{1}{2}\sqrt{t}x^{\frac{-1}{2}}$$

$$= \frac{y}{zt} + \frac{1}{2}\sqrt{\frac{t}{x}}$$

$$\frac{\partial f}{\partial y} = \frac{x}{zt} + \sqrt{z}$$

$$\frac{\partial f}{\partial z} = \frac{-xy}{tz^2} + \frac{y}{2\sqrt{z}}$$

$$\frac{\partial f}{\partial t} = \frac{-xy}{zt^2} + \frac{1}{2}\sqrt{\frac{x}{t}}$$

As a more complex example, consider $f(x, y) = e^{xy}\cos x \sin y$. The partial derivatives are listed next where we use the product rule in each case.

$$\frac{\partial f}{\partial x} = (ye^{xy}\sin y)(\cos x) + e^{xy}\sin y(-\sin x)$$

$$= e^{xy}\sin y(y\cos x - \sin x)$$

$$\frac{\partial f}{\partial y} = (xe^{xy}\cos x)(\sin y) + e^{xy}\cos x\cos y$$

$$= e^{xy}\cos x(x\sin y + \cos y)$$

Mixed Partial Derivatives

Just as with the derivative of a function of a single variable, we can take partial derivatives of a partial derivative. These are known as *mixed partials*. Additionally, we have more flexibility because we can change which variable we take the next partial derivative with respect to. For example, above, we saw that the partial derivative of $f(x,y) = \frac{xy}{zt} + y\sqrt{z} + \sqrt{xt}$ with respect to z is

$$\frac{\partial f}{\partial z} = \frac{-xy}{tz^2} + \frac{y}{2\sqrt{z}}$$

which is still a function of x, y, z, and t. Therefore, we can calculate second partial derivatives like so:

$$\frac{\partial^2 f}{\partial x\partial z} = \frac{-y}{tz^2}$$

$$\frac{\partial^2 f}{\partial y\partial z} = \frac{-x}{tz^2} + \frac{1}{2\sqrt{z}}$$

$$\frac{\partial^2 f}{\partial t\partial z} = \left(\frac{-xy}{z^2}\right)\left(\frac{-1}{t^2}\right)$$

$$= \frac{xy}{t^2z^2}$$

$$\frac{\partial^2 f}{\partial z^2} = \left(\frac{-xy}{t}\right)(-2)\left(\frac{1}{z^3}\right) + \left(\frac{y}{2}\right)\left(\frac{-1}{2}\right)\left(\frac{1}{z^{3/2}}\right)$$

$$= \frac{2xy}{tz^3} - \frac{y}{4z^{3/2}}$$

I'll explain the notation. We started with the partial derivative of f with respect to z, so we write $\partial f/\partial z$. Then, from this starting point, we are taking other partial derivatives. So, if we want to denote the partial with respect to x, we think of it this way:

$$\frac{\partial}{\partial x}\left(\frac{\partial f}{\partial z}\right) = \frac{\partial^2 f}{\partial x \partial z}$$

where we can think of the partial derivative operator "multiplying" the "numerator" and "denominator" like a fraction. To be clear, however, these are not fractions; the notation has just inherited the flavor of a fraction from its slope origins. Still, if the mnemonic is helpful, then it's helpful. For a second partial derivative, the variable it's taken with respect to is on the left. Also, if the variables are the same, an exponent (of sorts) is used, as in $\partial^2 f/\partial z^2$.

The Chain Rule for Partial Derivatives

To apply the chain rule to partial derivatives, we need to track all the variables. So, if we have $f(x, y)$, where both x and y are functions of other variables, $x(r, s)$ and $y(r, s)$, then we can find the partials of f with respect to r and s by applying the chain rule for each variable, x and y. Specifically,

$$\frac{\partial f}{\partial r} = \left(\frac{\partial f}{\partial x}\right)\left(\frac{\partial x}{\partial r}\right) + \left(\frac{\partial f}{\partial y}\right)\left(\frac{\partial y}{\partial r}\right)$$

$$\frac{\partial f}{\partial s} = \left(\frac{\partial f}{\partial x}\right)\left(\frac{\partial x}{\partial s}\right) + \left(\frac{\partial f}{\partial y}\right)\left(\frac{\partial y}{\partial s}\right)$$

As an example, let $f(x, y) = x^3 + y^3$ with $x(r, s) = 3r + 2s$ and $y(r, s) = r^2 - 3s$. Now find $\partial f/\partial r$ and $\partial f/\partial s$. To find these partials, we'll need to calculate six expressions,

$$\frac{\partial f}{\partial x} = 3x^2; \quad \frac{\partial f}{\partial y} = 3y^2; \quad \frac{\partial x}{\partial r} = 3; \quad \frac{\partial x}{\partial s} = 2; \quad \frac{\partial y}{\partial r} = 2r; \quad \frac{\partial y}{\partial s} = -3$$

so that the desired partials are

$$\frac{\partial f}{\partial r} = \left(\frac{\partial f}{\partial x}\right)\left(\frac{\partial x}{\partial r}\right) + \left(\frac{\partial f}{\partial y}\right)\left(\frac{\partial y}{\partial r}\right)$$

$$= (3x^2)(3) + (3y^2)(2r)$$

$$= 9x^2 + 6y^2r$$

$$= 9(3r + 2s)^2 + 6(r^2 - 3s)^2r$$

$$\frac{\partial f}{\partial s} = \left(\frac{\partial f}{\partial x}\right)\left(\frac{\partial x}{\partial s}\right) + \left(\frac{\partial f}{\partial y}\right)\left(\frac{\partial y}{\partial s}\right)$$

$$= (3x^2)(2) + (3y^2)(-3)$$

$$= 6x^2 - 9y^2$$

$$= 6(3r + 2s)^2 - 9(r^2 - 3s)^2$$

Just as with a function of a single variable, the chain rule for functions of more than one variable is recursive, such that if r and s were themselves functions of another variable, we could apply the chain rule one more time to find the partial of f with respect to that variable. For example, if we have $x(r, s)$, $y(r, s)$, with $r(w)$, $s(w)$, we find $\partial f/\partial w$ by using

$$\frac{\partial f}{\partial w} = \left(\frac{\partial f}{\partial x}\right)\left(\frac{\partial x}{\partial r}\right)\left(\frac{\partial r}{\partial w}\right) + \left(\frac{\partial f}{\partial y}\right)\left(\frac{\partial y}{\partial r}\right)\left(\frac{\partial r}{\partial w}\right) +$$

$$\left(\frac{\partial f}{\partial x}\right)\left(\frac{\partial x}{\partial s}\right)\left(\frac{\partial s}{\partial w}\right) + \left(\frac{\partial f}{\partial y}\right)\left(\frac{\partial y}{\partial s}\right)\left(\frac{\partial s}{\partial w}\right)$$

In the end, we need to remember that $\partial f/\partial w$ tells us how f will change for a small change in w. We'll use this fact during gradient descent.

This section concerned itself with the mechanical calculation of partial derivatives. Let's move on to explore more of the meaning behind these quantities. This will lead us to the idea of a gradient.

Gradients

In Chapter 8, we will dive into the matrix calculus representation we use in deep learning. However, before we do that, we will conclude this chapter by introducing the idea of a *gradient*. The gradient builds on the derivatives we've been calculating. In short, the gradient tells us how a function of more than one variable is changing and the direction in which it is changing the most.

Calculating the Gradient

If we have $f(x, y, z)$, we saw above how to calculate partial derivatives with respect to each of the variables. If we interpret the variables as positions on coordinate axes, we see that f is a function that returns a scalar, a single number, for any position in 3D space, (x, y, z). We could even go so far as to write $f(\boldsymbol{x})$ where $\boldsymbol{x} = (x, y, z)$ to acknowledge that f is a function of a vector input. As in previous chapters, we'll use lowercase bold letters to represent vectors, $\boldsymbol{x}$. Note, some people use $\vec{x}$ to represent vectors.

We can write vectors horizontally, as a row vector, like in the previous paragraph, or vertically,

$$\boldsymbol{x} = \begin{bmatrix} x \\ y \\ z \end{bmatrix}$$

as a column vector, where we've also used square brackets instead of parentheses. Either notation is acceptable. Unless we're intentionally sloppy, usually when we're discussing a vector in code, we'll assume that our vectors are column vectors. This means a vector is a matrix with n rows and one column, $n \times 1$.

A function that accepts a vector input and returns a single number as output is known as a *scalar field*. The canonical example of a scalar field is temperature. We can measure the temperature at any point in a room. We represent the location as a 3D vector relative to some chosen origin point, and the temperature is the value at that point, the average kinetic energy of the molecules in that region. We can also talk about functions that accept vectors as input and return a vector as output. These are known as *vector fields*. In both cases, the field part refers to the fact that, over some suitable domain, the function has a value for all inputs.

The gradient is the derivative of a function that accepts a vector as input. Mathematically, we represent the gradient as a generalization of the idea of partial derivatives to n dimensions. For example, in 3D space, we can write

$$\nabla f(\mathbf{x}) = \begin{bmatrix} \dfrac{\partial f}{\partial x} \\ \dfrac{\partial f}{\partial y} \\ \dfrac{\partial f}{\partial z} \end{bmatrix}$$

where the gradient operator, ∇, takes the partial derivative of f along each of its dimensions. The ∇ operator goes by multiple names, like *del*, *grad*, or *nabla*. We'll use ∇ and call it *del* when we're not simply saying "gradient operator."

In general, we can write

$$\mathbf{y} = \nabla f(\mathbf{x}) = \nabla f(x_0, x_1, \ldots, x_n) \equiv \begin{bmatrix} \dfrac{\partial f}{\partial x_0} \\ \dfrac{\partial f}{\partial x_1} \\ \dfrac{\partial f}{\partial x_2} \\ \vdots \\ \dfrac{\partial f}{\partial x_n} \end{bmatrix} \qquad (7.7)$$

Let's parse Equation 7.7. First, we have a function, f, that accepts a vector input, $\mathbf{x}$, and returns a scalar value. To this function, we apply the gradient operator:

$$\nabla = \begin{bmatrix} \dfrac{\partial}{\partial x_0} \\ \dfrac{\partial}{\partial x_1} \\ \dfrac{\partial}{\partial x_2} \\ \vdots \\ \dfrac{\partial}{\partial x_n} \end{bmatrix}$$

This returns a *vector* ($\mathbf{y}$). The gradient operator turns the scalar output of f into a vector. Let's spend some time thinking about what this means, what it's telling us about the value of the scalar field at a given position in space. (We'll use *space* when working with vectors, even if there's no meaningful way to visualize the space. An analogy to 3D space is helpful but only goes so far; mathematically, the idea of space is more general.)

As an example, consider a function in 2D space, $f(\boldsymbol{x}) = f(x, y) = x^2 + xy + y^2$. The gradient of f is then

$$\nabla f(\boldsymbol{x}) = \begin{bmatrix} \dfrac{\partial f}{\partial x} \\ \dfrac{\partial f}{\partial y} \end{bmatrix} = \begin{bmatrix} \dfrac{\partial}{\partial x}(x^2 + xy + y^2) \\ \dfrac{\partial}{\partial y}(x^2 + xy + y^2) \end{bmatrix} = \begin{bmatrix} 2x + y \\ 2y + x \end{bmatrix} \tag{7.8}$$

Since f is a scalar field, every point on the 2D plane has a function value. This is the output of $f(\boldsymbol{x}) = f(x, y)$. So, we can plot f in 3D to show us a surface changing with position. The gradient, however, gives us a set of equations. These equations collectively tell us the direction and magnitude of the change in the function value at a point, $\boldsymbol{x} = (x, y)$.

For a function of a single variable, there's only one slope at each point. Look again at the tangent line of Figure 7-1. At the point x_t, there's only one slope. The sign of the derivative at x_t gives the direction of the slope, and the absolute value of the derivative gives the magnitude (steepness) of the slope.

However, once we change to more than one dimension, we have a bit of a conundrum. Instead of only one slope tangent to the function, we now have an infinite number. We can imagine a line tangent to the function at some point and that the line points in any direction we so desire. The slope of the line tells us how the function value is changing in that particular direction. We can find the value of this change from the *directional derivative*, the dot product between the gradient at the point under consideration and a unit vector in the direction we're interested in:

$$D_{\boldsymbol{u}}f(\boldsymbol{x}) \equiv \boldsymbol{u} \cdot \nabla f(\boldsymbol{x}) = \boldsymbol{u}^T \nabla f(\boldsymbol{x}) = \|\boldsymbol{u}\| \|\nabla f(\boldsymbol{x})\| \cos\theta$$

where $\boldsymbol{u}$ is a unit vector in a particular direction, $\nabla f(\boldsymbol{x})$ is the gradient of the function at the point $\boldsymbol{x}$, and θ is the angle between them. The directional derivative is maximized when $\cos\theta$ is maximized, and this happens at $\theta = 0$. Therefore, the direction of the maximum change in a function at any point is the gradient at that point.

As an example, let's pick a point on the 2D plane, say $\boldsymbol{x} = (x, y) = (0.5, -0.4)$ with $f(x, y) = x^2 + xy + y^2$ from above. Then, the function value at $\boldsymbol{x}$ is $x^2 + xy + y^2 = (0.5)^2 + (0.5)(-0.4) + (-0.4)^2 = 0.21$, a scalar. However, the gradient at this point is

$$\nabla f(0.5, -0.4) = \begin{bmatrix} 2x + y = 2(0.5) + (-0.4) = 0.6 \\ 2y + x = 2(-0.4) + (0.5) = -0.3 \end{bmatrix}$$

Therefore, we now know that at the point $(0.5, -0.4)$, the direction of the largest change in f is in the direction $(0.6, -0.3)$ and has a magnitude of $\sqrt{(0.6)^2 + (-0.3)^2} \approx 0.67$.

Visualizing the Gradient

Let's make all of this less abstract. The top part of Figure 7-4 shows a plot of $f(x, y) = x^2 + xy + y^2$ at selected points.

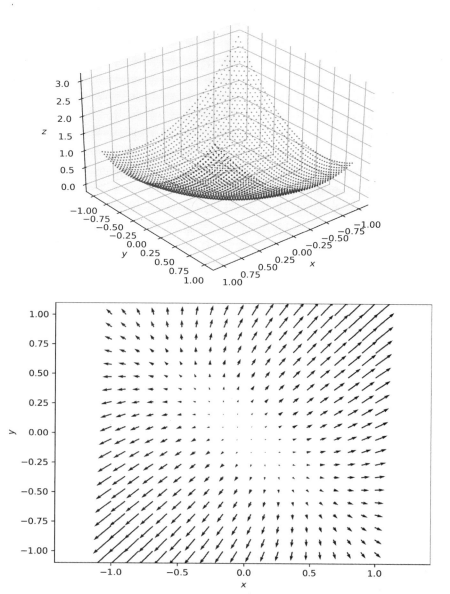

Figure 7-4: A plot of $x^2 + xy + y^2$ (top) and a 2D projection of the associated gradient field (bottom)

The code to generate this plot is straightforward:

```
import numpy as np
x = np.linspace(-1.0,1.0,50)
```

```
y = np.linspace(-1.0,1.0,50)
xx = []; yy = []; zz = []
for i in range(50):
    for j in range(50):
        xx.append(x[i])
        yy.append(y[j])
        zz.append(x[i]*x[i]+x[i]*y[j]+y[j]*y[j])
x = np.array(xx)
y = np.array(yy)
z = np.array(zz)
```

Here, we explicitly loop to generate the set of scatter plot points, x, y, and z, to clearly show what's happening. First, we use NumPy to generate vectors of 50 evenly spaced points $[-1, 1]$ in x and y. Then we set up a double loop so that each x gets paired with each y to calculate the function value, z. Temporary lists xx, yy, and zz hold the triplets. Finally, we convert the lists to NumPy arrays for plotting.

The code to generate the scatter plot is

```
from mpl_toolkits.mplot3d import Axes3D
import matplotlib.pylab as plt
fig = plt.figure()
ax = fig.add_subplot(111, projection='3d')
ax.scatter(x, y, z, marker='.', s=2, color='b')
ax.view_init(30,20)
plt.draw()
plt.show()
```

We first load the `matplotlib` extension for 3D plotting, and then we set the subplot for a 3D projection. The plot itself is made with `ax.scatter`, while `ax.view_init` and `plt.draw` rotate the plot to give us a better view of the shape of the function before showing it.

In the bottom part of Figure 7-4, we see a vector plot of the gradient field for $x^2 + xy + y^2$. This plot shows the direction and relative magnitude of the gradient vector at a grid of points, (x, y). Recall, the gradient is a vector field, so each point on the xy-plane has an associated vector pointing in the direction of the greatest change in the function value. Mentally, we can see how the vector plot relates to the function plot in the top part of Figure 7-4, where function values near $(-1, -1)$ and $(1, 1)$ are changing quickly, whereas for points near $(0, 0)$ they're changing slowly.

The code to generate the vector field plot is

```
fig = plt.figure()
ax = fig.add_subplot(111)
x = np.linspace(-1.0,1.0,20)
y = np.linspace(-1.0,1.0,20)
xv, yv = np.meshgrid(x, y, indexing='ij', sparse=False)
dx = 2*xv + yv
dy = 2*yv + xv
```

```
ax.quiver(xv, yv, dx, dy, color='b')
plt.axis('equal')
plt.show()
```

We first define the figure (fig) and subplot for 2D (no projection key-word). Then, we need a grid of points. Above, we looped to get this grid so we could understand what needed to be generated. Here, we use NumPy to generate the grid for us via np.meshgrid. Note, we pass np.meshgrid the same x and y vectors we had above to define the domain.

The next two lines are a direct implementation of the gradient of f, Equation 7.8. These are the vectors we want to plot, with dx and dy giving us the direction and magnitude, while xv and yv are the set of input points—400 total.

The plot uses ax.quiver (since it plots arrows). The arguments are the grid of points (xv, yv) and associated x and y values of the vectors at those points (dx, dy). Finally, we ensure the axes are equal (plt.axis) to avoid warp-ing the vector display, then show the plot.

We'll conclude our introduction of gradients here. We'll see them again throughout the remainder of the book, in the notation in Chapter 8 and the gradient descent discussions of Chapter 11.

Summary

This chapter introduced the main concepts of differential calculus. We start-ed with the notion of slope and learned the difference between secant and tangent lines for a function of a single variable. We then formally defined the derivative as the slope of a secant line as it approaches a single point. From there, we learned the basic rules of differentiation and saw how to ap-ply them.

Next, we learned about the minima and maxima of a function and how to find these points using derivatives. We then introduced partial derivatives as a way to calculate derivatives for functions of more than one variable. Par-tial derivatives then led us to the gradient, which turns a scalar field into a vector field and tells us the direction in which the function is changing the most. We calculated an example gradient in 2D and saw how to generate plots showing the relationship between the function and the gradient. We learned the crucial fact that the gradient of a function points in the direc-tion of the maximum change in the function value at a point.

Let's continue our exploration of the math behind deep learning and move into the world of matrix calculus.

8

MATRIX CALCULUS

Chapter 7 introduced us to differential calculus. In this chapter, we'll discuss *matrix calculus*, which extends differentiation to functions involving vectors and matrices.

Deep learning works extensively with vectors and matrices, so it makes sense to develop a notation and approach to representing derivatives involving these objects. That's what matrix calculus gives us. We saw a hint of this at the end of Chapter 7, when we introduced the gradient to represent the derivative of a scalar function of a vector—a function that accepts a vector argument and returns a scalar, $f(x)$.

We'll start with the table of matrix calculus derivatives and their definitions. Next, we'll examine some identities involving matrix derivatives. Mathematicians love identities; however, to preserve our sanity, we'll only consider a handful. Some special matrices come out of matrix calculus, namely the Jacobian and Hessian. You'll run into both of these matrices during your sojourn through deep learning, so we'll consider them next in the context of optimization. Recall that training a neural network is, fundamentally, an optimization problem, so understanding what these special matrices represent and how we use them is especially important. We'll close the chapter with some examples of matrix derivatives.

The Formulas

Table 8-1 summarizes the matrix calculus derivatives we'll explore in this chapter. These are the ones commonly used in practice.

Table 8-1: Matrix Calculus Derivatives

	Scalar	Vector	Matrix
Scalar	$\partial f/\partial x$	$\partial f/\partial x$	$\partial F/\partial x$
Vector	$\partial f/\partial x$	$\partial f/\partial x$	—
Matrix	$\partial f/\partial X$	—	—

The columns of Table 8-1 represent the function, meaning the return value. Notice we use three versions of the letter f: regular, bold, and capital. We use f if the return value is a scalar, f if a vector, and F if a matrix. The rows of Table 8-1 are the variables the derivatives are calculated with respect to. The same notation applies: x is a scalar, x is a vector, and X is a matrix.

Table 8-1 defines six derivatives, but there are nine cells in the table. While possible to define, the remaining derivatives are not standardized or used often enough to make covering them worthwhile. That's good for us, as the six are enough of a challenge for our mathematical brains.

The first derivative in Table 8-1, the one in the upper left, is the normal derivative of Chapter 7, a function producing a scalar with respect to a scalar. (Refer to Chapter 7 for everything you need to know about standard differentiation.)

We'll cover the remaining five derivatives in the sections below. We define each one in terms of scalar derivatives. We'll first show the definition and then explain what the notation represents. The definition will help you build a model in your head of what the derivative is. I suspect that by the end of this section you'll be predicting the definitions in advance.

Before we start, however, there is a complication we should discuss. Matrix calculus is notation heavy, but there's no universal agreement on the notation. We've seen this before with the many ways to indicate differentiation. For matrix calculus, the two approaches are *numerator layout* or *denominator layout*. Specific disciplines seem to favor one over the other, though exceptions are almost the norm, as is mixing notations. For deep learning, a nonscientific perusal on my part seems to indicate a slight preference for numerator layout, so that's what we'll use here. Just be aware that there are two forms out there. One is typically the transpose of the other.

A Vector Function by a Scalar Argument

A vector function accepting a scalar argument is our first derivative; see Table 8-1, second column of the first row. We write such a function as $f(x)$

to indicate a scalar argument, x, and a vector output, f. Functions like f take a scalar and map it to a multidimensional vector:

$$f : \mathbb{R} \to \mathbb{R}^m$$

Here, m is the number of elements in the output vector. Functions like f are known as *vector-valued functions* with scalar arguments.

A parametric curve in 3D space is an excellent example of such a function. Those functions are often written as

$$f(x) = f_0(x)\hat{\mathbf{x}} + f_1(x)\hat{\mathbf{y}} + f_2(x)\hat{\mathbf{z}}$$

where $\hat{\mathbf{x}}$, $\hat{\mathbf{y}}$, and $\hat{\mathbf{z}}$ are unit vectors in the x, y, and z directions.

Figure 8-1 shows a plot of a 3D parametric curve,

$$f(t) = t\cos(t)\hat{\mathbf{x}} + t\sin(t)\hat{\mathbf{y}} + t\hat{\mathbf{z}} \tag{8.1}$$

where, as t varies, the three axis values also vary to trace out the spiral. Here, each value of t specifies a single point in 3D space.

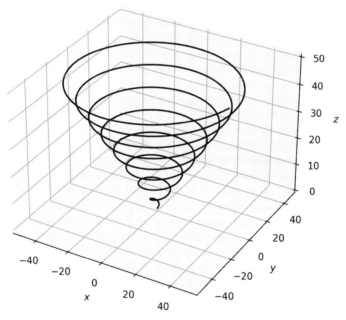

Figure 8-1: A 3D parametric curve

In matrix calculus notation, we don't write f as shown in Equation 8.1. Instead, we write f as a column vector of the functions,

$$f(t) = \begin{bmatrix} t\cos(t) \\ t\sin(t) \\ t \end{bmatrix}$$

and, in general,

$$f(x) = \begin{bmatrix} f_0(x) \\ f_1(x) \\ \vdots \\ f_{n-1}(x) \end{bmatrix}$$

for f with n elements.

The derivative of $f(x)$ is known as the *tangent vector*. What does the derivative look like? Since f is a vector, we might expect the derivative of f to be the derivatives of the functions representing each element of f, and we'd be right:

$$\frac{\partial f}{\partial x} = \begin{bmatrix} \partial f_0/\partial x \\ \partial f_1/\partial x \\ \vdots \\ \partial f_{n-1}/\partial x \end{bmatrix}$$

Let's look at a simple example. First, we will define $f(x)$, and then the derivative:

$$f(x) = \begin{bmatrix} 2x^2 - 3x + 2 \\ x^3 - 3 \end{bmatrix}, \ \frac{\partial f}{\partial x} = \begin{bmatrix} 4x - 3 \\ 3x^2 \end{bmatrix}$$

Here, each element of $\partial f/\partial x$ is the derivative of the corresponding function in f.

A Scalar Function by a Vector Argument

In Chapter 7, we learned that a function accepting a vector input but returning a scalar is a scalar field:

$$f : \mathbb{R}^m \to \mathbb{R}$$

We also learned that the derivative of this function is the gradient. In matrix calculus notation, we write $\partial f/\partial x$ for $f(x)$ as

$$\frac{\partial f}{\partial x} = \begin{bmatrix} \dfrac{\partial f}{\partial x_0} & \dfrac{\partial f}{\partial x_1} & \cdots & \dfrac{\partial f}{\partial x_{m-1}} \end{bmatrix}$$

where $x = \begin{bmatrix} x_0 & x_1 & \cdots & x_{m-1} \end{bmatrix}^\top$ is a vector of variables, and f is a function of those variables.

Notice, because we decided to use the numerator layout approach, $\partial f/\partial \boldsymbol{x}$ is written as a *row* vector. So, to be true to our notation, we have to write

$$\nabla f(\boldsymbol{x}) = \left[\frac{\partial f}{\partial \boldsymbol{x}}\right]^{\top}$$

turning the row vector into a column vector to match the gradient. Remember that ∇ is the symbol for gradient; we saw an example of the gradient in Chapter 7.

A Vector Function by a Vector

If the derivative of a vector-valued function by a scalar produces a column vector, and the derivative of a scalar function by a vector results in a row vector, does the derivative of a vector-valued function by a vector produce a matrix? The answer is yes. In this case, we're contemplating $\partial f/\partial \boldsymbol{x}$ for $f(\boldsymbol{x})$, a function that accepts a vector input and returns a vector.

The numerator layout convention gave us a column vector for the derivative of $f(x)$, implying we need a row for each of the functions in f. Similarly, the derivative of $f(\boldsymbol{x})$ produced a row vector. Therefore, merging the two gives us the derivative of $f(\boldsymbol{x})$:

$$\frac{\partial f}{\partial \boldsymbol{x}} = \begin{bmatrix} \frac{\partial f_0}{\partial x_0} & \frac{\partial f_0}{\partial x_1} & \cdots & \frac{\partial f_0}{\partial x_{m-1}} \\ \frac{\partial f_1}{\partial x_0} & \frac{\partial f_1}{\partial x_1} & \cdots & \frac{\partial f_1}{\partial x_{m-1}} \\ \vdots & \vdots & & \vdots \\ \frac{\partial f_{n-1}}{\partial x_0} & \frac{\partial f_{n-1}}{\partial x_1} & \cdots & \frac{\partial f_{n-1}}{\partial x_{m-1}} \end{bmatrix} \tag{8.2}$$

This is for a function, f, returning an n-element vector and accepting an m-element vector, $\boldsymbol{x}$, as its argument:

$$f: \mathbb{R}^m \to \mathbb{R}^n$$

Each row of f is a scalar function of $\boldsymbol{x}$, for example, $f_0(\boldsymbol{x})$. Therefore, we can write Equation 8.2 as

$$\frac{\partial f}{\partial \boldsymbol{x}} = \begin{bmatrix} \nabla f_0(\boldsymbol{x})^{\top} \\ \nabla f_1(\boldsymbol{x})^{\top} \\ \vdots \\ \nabla f_{n-1}(\boldsymbol{x})^{\top} \end{bmatrix} \tag{8.3}$$

This gives us the matrix as a collection of gradients, one for each scalar function in f. We'll return to this matrix later in the chapter.

A Matrix Function by a Scalar

If $f(x)$ is a function accepting a scalar argument but returning a vector, then we'd be correct in assuming $F(x)$ can be thought of as a function accepting a scalar argument but returning a matrix:

$$F : \mathbb{R} \to \mathbb{R}^{n \times m}$$

For example, assume F is an $n \times m$ matrix of scalar functions:

$$F = \begin{bmatrix} f_{00}(x) & f_{01}(x) & \cdots & f_{0,m-1}(x) \\ f_{10}(x) & f_{11}(x) & \cdots & f_{1,m-1}(x) \\ \vdots & \vdots & & \vdots \\ f_{n-1,0}(x) & f_{n-1,1}(x) & \cdots & f_{n-1,m-1}(x) \end{bmatrix}$$

The derivative with respect to the argument, x, is straightforward:

$$\frac{\partial F}{\partial x} = \begin{bmatrix} \frac{\partial f_{00}}{\partial x} & \frac{\partial f_{01}}{\partial x} & \cdots & \frac{\partial f_{0,m-1}}{\partial x} \\ \frac{\partial f_{10}}{\partial x} & \frac{\partial f_{11}}{\partial x} & \cdots & \frac{\partial f_{1,m-1}}{\partial x} \\ \vdots & \vdots & & \vdots \\ \frac{\partial f_{n-1,0}}{\partial x} & \frac{\partial f_{n-1,1}}{\partial x} & \cdots & \frac{\partial f_{n-1,m-1}}{\partial x} \end{bmatrix}$$

As we saw above, the derivative of a vector-valued function by a scalar is called the tangent vector. By analogy, then, the derivative of a matrix-valued function by a scalar is the *tangent matrix*.

A Scalar Function by a Matrix

Now let's consider $f(X)$, a function accepting a matrix and returning a scalar:

$$f : \mathbb{R}^{n \times m} \to \mathbb{R}$$

We'd be correct in thinking that the derivative of f with respect to the matrix, X, is itself a matrix. However, to be true to our numerator layout convention, the resulting matrix is not arranged like X, but instead like $X^\top$, the transpose of X.

Why use the transpose of X instead of X itself? To answer the question, we need to look back to how we defined $\partial f/\partial x$. There, even though x is a column vector, according to standard convention, we said the derivative is a *row* vector. We used $x^\top$ as the ordering. Therefore, to be consistent, we need to arrange $\partial f/\partial X$ by the transpose of X and change the columns of X into rows in the derivative. As a result, we have the following definition.

$$\frac{\partial f}{\partial \mathbf{X}} = \begin{bmatrix} \frac{\partial f}{\partial x_{00}} & \frac{\partial f}{\partial x_{10}} & \cdots & \frac{\partial f}{\partial x_{n-1,0}} \\ \frac{\partial f}{\partial x_{01}} & \frac{\partial f}{\partial x_{11}} & \cdots & \frac{\partial f}{\partial x_{n-1,1}} \\ \vdots & \vdots & & \vdots \\ \frac{\partial f}{\partial x_{0,m-1}} & \frac{\partial f}{\partial x_{1,m-1}} & \cdots & \frac{\partial f}{\partial x_{n-1,m-1}} \end{bmatrix} \tag{8.4}$$

This is an $m \times n$ output matrix for the $n \times m$ input matrix, $\mathbf{X}$.

Equation 8.4 defines the *gradient matrix*, which, for matrices, plays a role similar to that of the gradient, $\nabla f(\mathbf{x})$. Equation 8.4 also completes our collection of matrix calculus derivatives. Let's move on to consider some matrix derivative identities.

The Identities

Matrix calculus involves scalars, vectors, matrices, and functions thereof, which themselves return scalars, vectors, or matrices, implying many identities and relationships exist. However, here we'll concentrate on basic identities showing the relationship between matrix calculus and the differential calculus of Chapter 7.

Each of the following subsections presents identities related to the specific type of derivative indicated. The identities cover fundamental relationships and, when applicable, the chain rule. In all cases, the results follow the numerator layout scheme we've used throughout the chapter.

A Scalar Function by a Vector

We begin with identities related to a scalar function with a vector input. If not otherwise specified, f and g are functions of a vector, $\mathbf{x}$, and return a scalar. A constant vector that doesn't depend on $\mathbf{x}$ is given as $\mathbf{a}$, and a denotes a scalar constant.

The basic rules are intuitive:

$$\frac{\partial}{\partial \mathbf{x}}(af) = a\frac{\partial f}{\partial \mathbf{x}} \tag{8.5}$$

and

$$\frac{\partial}{\partial \mathbf{x}}(f+g) = \frac{\partial f}{\partial \mathbf{x}} + \frac{\partial g}{\partial \mathbf{x}} \tag{8.6}$$

These show that multiplication by a scalar constant acts as it did in Chapter 7, as does the linearity of the partial derivative.

The product rule also works as we expect:

$$\frac{\partial}{\partial \boldsymbol{x}}(fg) = f\frac{\partial g}{\partial \boldsymbol{x}} + g\frac{\partial f}{\partial \boldsymbol{x}} \tag{8.7}$$

Let's pause here and remind ourselves of the inputs and outputs for the equations above. We know that the derivative of a scalar function by a vector argument is a row vector in our notation. So, Equation 8.5 returns a row vector multiplied by a scalar—each element of the derivative is multiplied by a.

As differentiation is a linear operator, it distributes over addition, so Equation 8.6 delivers two terms, each a row vector generated by the respective derivative.

For Equation 8.7, the product rule, the result again includes two terms. In each case, the derivative returns a row vector, which is multiplied by a scalar function value, either $f(\boldsymbol{x})$ or $g(\boldsymbol{x})$. Therefore, the output of Equation 8.7 is also a row vector.

The scalar-by-vector chain rule becomes

$$\frac{\partial}{\partial \boldsymbol{x}}f(g) = \frac{\partial f}{\partial g}\frac{\partial g}{\partial \boldsymbol{x}} \tag{8.8}$$

where $f(g)$ returns a scalar and accepts a scalar argument, while $g(\boldsymbol{x})$ returns a scalar and accepts a vector argument. The end result is a row vector. Let's work through a complete example to demonstrate.

We have a vector, $\boldsymbol{x} = [x_0, x_1, x_2]^\top$; a function of that vector written in component form, $g(\boldsymbol{x}) = x_0 + x_1 x_2$; and a function of g, $f(g) = g^2$. According to Equation 8.8, the derivative of f with respect to $\boldsymbol{x}$ is

$$\frac{\partial f}{\partial \boldsymbol{x}} = \frac{\partial f}{\partial g}\frac{\partial g}{\partial \boldsymbol{x}}$$

$$= \frac{\partial f}{\partial g}\left[\begin{array}{ccc} \dfrac{\partial g}{\partial x_0} & \dfrac{\partial g}{\partial x_1} & \dfrac{\partial g}{\partial x_2} \end{array}\right]$$

$$= (2g)\begin{bmatrix} 1 & x_2 & x_1 \end{bmatrix}$$

$$= \begin{bmatrix} 2g & 2gx_2 & 2gx_1 \end{bmatrix}$$

$$= \begin{bmatrix} 2(x_0 + x_1 x_2) & 2(x_0 + x_1 x_2)x_2 & 2(x_0 + x_1 x_2)x_1 \end{bmatrix}$$

$$= \begin{bmatrix} 2x_0 + 2x_1 x_2 & 2x_0 x_2 + 2x_1 x_2^2 & 2x_0 x_1 + 2x_1^2 x_2 \end{bmatrix}$$

To check our result, we can work through from $g(x) = x_0 + x_1 x_2$ and $f(g) = g^2$ to find $f(x)$ directly via substitution. Doing this gives us

$$f(x) = x_0^2 + 2x_0 x_1 x_2 + x_1^2 x_2^2$$

from which we get

$$\frac{\partial f}{\partial x} = \begin{bmatrix} \dfrac{\partial f}{\partial x_0} & \dfrac{\partial f}{\partial x_1} & \dfrac{\partial f}{\partial x_2} \end{bmatrix}$$

$$= \begin{bmatrix} 2x_0 + 2x_1 x_2 & 2x_0 x_2 + 2x_1 x_2^2 & 2x_0 x_1 + 2x_1^2 x_2 \end{bmatrix}$$

which matches the result we found using the chain rule. Of course, in this simple example, it was easier to work through the substitution before taking the derivative, but we proved our case all the same.

We're not entirely through with scalar-by-vector identities, however. The dot product takes two vectors and produces a scalar, so it fits with the functional form we're working with, even though the arguments to the dot product are vectors.

For example, consider this result:

$$\frac{\partial}{\partial x}(a \cdot x) = \frac{\partial}{\partial x}(a^\top x) = a^\top \tag{8.9}$$

Here, we have the derivative of the dot product between x and a vector a that does not depend on x.

We can expand on Equation 8.9, replacing x with a vector-valued function, $f(x)$:

$$\frac{\partial}{\partial x}(a \cdot f) = \frac{\partial}{\partial x}(a^\top f) = a^\top \frac{\partial f}{\partial x} \tag{8.10}$$

What's the form of this result? Assume f accepts an m-element input and returns an n-element vector output. Likewise, assume a to be an n-element vector. From Equation 8.2, we know the derivative $\partial f / \partial x$ to be an $n \times m$ matrix. Therefore, the final result is a $(1 \times n) \times (n \times m) \to 1 \times m$ row vector. Good! We know the derivative of a scalar function by a vector should be a row vector when using the numerator layout convention.

Finally, the derivative of the dot product of two vector-valued functions, f and g, is

$$\frac{\partial}{\partial x}(f \cdot g) = \frac{\partial}{\partial x}(f^\top g) = f^\top \frac{\partial g}{\partial x} + g^\top \frac{\partial f}{\partial x} \tag{8.11}$$

If Equation 8.10 is a row vector, then the sum of two terms like it is also a row vector.

A Vector Function by a Scalar

Vector-by-scalar differentiation, Table 8-1, first row, second column, is less common in machine learning, so we'll only examine a few identities. The first are multiplications by constants:

$$\frac{\partial}{\partial x}(af) = a\frac{\partial f}{\partial x}$$

$$\frac{\partial}{\partial x}(Af) = A\frac{\partial f}{\partial x}$$

Note, we can multiply on the left by a matrix, as the derivative is a column vector.

The sum rule still applies,

$$\frac{\partial}{\partial x}(f+g) = \frac{\partial f}{\partial x} + \frac{\partial g}{\partial x}$$

as does the chain rule,

$$\frac{\partial}{\partial x}(f(g)) = \frac{\partial f}{\partial g}\frac{\partial g}{\partial x} \qquad (8.12)$$

Equation 8.12 is correct, since the derivative of a vector by a scalar is a column vector, and the derivative of a vector by a vector is a matrix. Therefore, multiplying the matrix on the right by a column vector returns a column vector, as expected.

Two other derivatives involving dot products with respect to a scalar are worth knowing about. The first is similar to Equation 8.11 but with two vector-valued functions of a scalar:

$$\frac{\partial}{\partial x}(f \cdot g) = f \cdot \frac{\partial g}{\partial x} + \frac{\partial f}{\partial x} \cdot g$$

$$= f^\top \frac{\partial g}{\partial x} + g^\top \frac{\partial f}{\partial x}$$

The second derivative concerns the composition of $f(g)$ and $g(x)$ with respect to x:

$$\frac{\partial}{\partial x}(f(g)) = \frac{\partial f}{\partial g} \cdot \frac{\partial g}{\partial x} = \frac{\partial f}{\partial g}\frac{\partial g}{\partial x}$$

which is the dot product of a row vector and a column vector.

A Vector Function by a Vector

The derivatives of vector-valued functions with vector arguments are common in physics and engineering. In machine learning, they show up during backpropagation, for example, at the derivative of the loss function. Let's begin with some straightforward identities:

$$\frac{\partial}{\partial \boldsymbol{x}}(a\boldsymbol{f}) = a\frac{\partial \boldsymbol{f}}{\partial \boldsymbol{x}}$$

$$\frac{\partial}{\partial \boldsymbol{x}}(A\boldsymbol{f}) = A\frac{\partial \boldsymbol{f}}{\partial \boldsymbol{x}}$$

and

$$\frac{\partial}{\partial \boldsymbol{x}}(\boldsymbol{f} + \boldsymbol{g}) = \frac{\partial \boldsymbol{f}}{\partial \boldsymbol{x}} + \frac{\partial \boldsymbol{g}}{\partial \boldsymbol{x}}$$

where the result is the sum of two matrices.

The chain rule is next and works as it did above for scalar-by-vector and vector-by-scalar derivatives:

$$\frac{\partial}{\partial \boldsymbol{x}}(\boldsymbol{f}(\boldsymbol{g})) = \frac{\partial \boldsymbol{f}}{\partial \boldsymbol{g}}\frac{\partial \boldsymbol{g}}{\partial \boldsymbol{x}}$$

with the result being the product of two matrices.

A Scalar Function by a Matrix

For functions of a matrix, $\boldsymbol{X}$, returning a scalar, we have the usual form for the sum rule:

$$\frac{\partial}{\partial \boldsymbol{X}}(f + g) = \frac{\partial f}{\partial \boldsymbol{X}} + \frac{\partial g}{\partial \boldsymbol{X}}$$

with the result being the sum of two matrices. Recall, if $\boldsymbol{X}$ is an $n \times m$ matrix, the derivative in numerator layout notation is an $m \times n$ matrix.

The product rule also works as expected:

$$\frac{\partial}{\partial \boldsymbol{X}}(fg) = f\frac{\partial g}{\partial \boldsymbol{X}} + g\frac{\partial f}{\partial \boldsymbol{X}}$$

However, the chain rule is different. It depends on $f(g)$, a scalar function accepting a scalar input, and $g(\boldsymbol{X})$, a scalar function accepting a matrix input. With this restriction, the form of the chain rule looks familiar:

$$\frac{\partial}{\partial \boldsymbol{X}}(f(g)) = \frac{\partial f}{\partial g}\frac{\partial g}{\partial \boldsymbol{X}} \tag{8.13}$$

Let's see Equation 8.13 in action. First, we need X, a 2×2 matrix:

$$X = \begin{bmatrix} x_0 & x_1 \\ x_2 & x_3 \end{bmatrix}$$

Next, we need $f(g) = \frac{1}{2}g^2$ and $g(X) = x_0 x_3 + x_1 x_2$. Notice, while $g(X)$ accepts a matrix input, the result is a scalar calculated from the values of the matrix.

To apply the chain rule, we need two derivatives,

$$\frac{\partial f}{\partial g} = g$$

$$\frac{\partial g}{\partial X} = \begin{bmatrix} \partial g / \partial x_0 & \partial g / \partial x_2 \\ \partial g / \partial x_1 & \partial g / \partial x_3 \end{bmatrix} = \begin{bmatrix} x_3 & x_1 \\ x_2 & x_0 \end{bmatrix}$$

where we are again using numerator layout for the result.

To find the overall result, we calculate

$$\frac{\partial f}{\partial X} = \frac{\partial f}{\partial g}\frac{\partial g}{\partial X}$$

$$= g \begin{bmatrix} x_3 & x_1 \\ x_2 & x_0 \end{bmatrix}$$

$$= \begin{bmatrix} x_3 g & x_1 g \\ x_2 g & x_0 g \end{bmatrix}$$

$$= \begin{bmatrix} x_3(x_0 x_3 + x_1 x_2) & x_1(x_0 x_3 + x_1 x_2) \\ x_2(x_0 x_3 + x_1 x_2) & x_0(x_0 x_3 + x_1 x_2) \end{bmatrix}$$

To check, we combine the functions to write a single function, $f(X) = \frac{1}{2}(x_0 x_3 + x_1 x_2)^2$, and calculate the derivative using the standard chain rule for each element of the resulting matrix. This gives us

$$\frac{\partial f}{\partial X} = \begin{bmatrix} \partial f / \partial x_0 & \partial f / \partial x_2 \\ \partial f / \partial x_1 & \partial f / \partial x_3 \end{bmatrix} = \begin{bmatrix} x_3(x_0 x_3 + x_1 x_2) & x_1(x_0 x_3 + x_1 x_2) \\ x_2(x_0 x_3 + x_1 x_2) & x_0(x_0 x_3 + x_1 x_2) \end{bmatrix}$$

matching the previous result.

We have our definitions and identities. Let's revisit the derivative of a vector-valued function with a vector argument, as the resulting matrix is special. We'll encounter it frequently in deep learning.

Jacobians and Hessians

Equation 8.2 defined the derivative of a vector-valued function, f, with respect to a vector, x:

$$J_x = \frac{\partial f}{\partial x} = \begin{bmatrix} \frac{\partial f_0}{\partial x_0} & \frac{\partial f_0}{\partial x_1} & \cdots & \frac{\partial f_0}{\partial x_{m-1}} \\ \frac{\partial f_1}{\partial x_0} & \frac{\partial f_1}{\partial x_1} & \cdots & \frac{\partial f_1}{\partial x_{m-1}} \\ \vdots & \vdots & & \vdots \\ \frac{\partial f_{n-1}}{\partial x_0} & \frac{\partial f_{n-1}}{\partial x_1} & \cdots & \frac{\partial f_{n-1}}{\partial x_{m-1}} \end{bmatrix} \tag{8.14}$$

This derivative is known as the *Jacobian matrix, J*, or simply the *Jacobian*, and you'll encounter it from time to time in the deep learning literature, especially during discussions of gradient descent and other optimization algorithms used in training models. The Jacobian sometimes has a subscript to indicate the variable it is with respect to; for example, J_x if with respect to x. When the context is clear, we'll often neglect the subscript.

In this section, we'll discuss the Jacobian and what it means. Then we'll introduce another matrix, the *Hessian matrix* (or just the *Hessian*), which is based on the Jacobian, and learn how to use it in optimization problems.

The essence of this section is the following: the Jacobian is the generalization of the first derivative, and the Hessian is the generalization of the second derivative.

Concerning Jacobians

We saw previously that we can think of Equation 8.14 as a stack of transposed gradient vectors (Equation 8.3):

$$J_x = \begin{bmatrix} \nabla f_0(x)^\top \\ \nabla f_1(x)^\top \\ \vdots \\ \nabla f_{n-1}(x)^\top \end{bmatrix}$$

Viewing the Jacobian as a stack of gradients gives us a clue as to what it represents. Recall, the gradient of a scalar field, a function accepting a vector argument and returning a scalar, points in the direction of the maximum change in the function. Similarly, the Jacobian gives us information about how the vector-valued function behaves in the vicinity of some point, x_p. The Jacobian is to vector-valued functions of vectors what the gradient is to scalar-valued functions of vectors; it tells us about how the function changes for a small change in the position of x_p.

One way to think of the Jacobian is as a generalization of the more specific cases we encountered in Chapter 7. Table 8-2 shows the relationship between the function and what its derivative measures.

Table 8-2: The Relationship Between Jacobians, Gradients, and Slopes

Function	Derivative
f(x)	∂**f**/∂**x**, Jacobian matrix
f(**x**)	∂f/∂**x**, gradient vector
f(x)	df/dx, slope

The Jacobian matrix is the most general of the three. If we limit the function to a scalar, then the Jacobian matrix becomes the gradient vector (row vector in numerator layout). If we limit the function and argument to scalars, the gradient becomes the slope. In a sense, they all indicate the same thing: how the function is changing around a point in space.

The Jacobian has many uses. I'll present two examples. The first is from systems of differential equations. The second uses Newton's method to find the roots of a vector-valued function. We'll see Jacobians again when we discuss backpropagation, as that requires calculating derivatives of a vector-valued function with respect to a vector.

Autonomous Differential Equations

A *differential equation* combines derivatives and function values in one equation. Differential equations show up everywhere in physics and engineering. Our example comes from the theory of *autonomous systems*, which are differential equations where the independent variable does not appear on the right-hand side. For instance, if the system consists of values of the function and first derivatives with respect to time, t, there is no t explicit in the equations governing the system.

The previous paragraph is just for background; you don't need to memorize it. Working with systems of autonomous differential equations ultimately leads to the Jacobian, which is our goal. We can view the system as a vector-valued function, and we'll use the Jacobian to characterize the critical points of that system (the points where the derivative is zero). We worked with critical points of functions in Chapter 7.

For example, let's explore the following system of equations:

$$\frac{dx}{dt} = 4x - 2xy$$

$$\frac{dy}{dt} = 2y + xy - 2y^2$$

This system includes two functions, $x(t)$ and $y(t)$, and they are coupled so that the rate of change of $x(t)$ depends on the value of x and the value of y, and vice versa.

We'll view the system as a single, vector-valued function:

$$f(x) = \begin{bmatrix} f_0 \\ f_1 \end{bmatrix} = \begin{bmatrix} 4x_0 - 2x_0x_1 \\ 2x_1 + x_0x_1 - 2x_1^2 \end{bmatrix}, \quad x = \begin{bmatrix} x_0 \\ x_1 \end{bmatrix} \tag{8.15}$$

where we replace x with x_0 and y with x_1.

The system that f represents has critical points at locations where $f = 0$, with 0 being the 2×1 dimensional zero vector. The critical points are

$$c_0 = \begin{bmatrix} 0 \\ 0 \end{bmatrix}, \quad c_1 = \begin{bmatrix} 0 \\ 1 \end{bmatrix}, \quad c_2 = \begin{bmatrix} 2 \\ 2 \end{bmatrix} \tag{8.16}$$

where substitution into f shows that each point returns the zero vector. For the time being, assume we were given the critical points, and now we want to characterize them.

To characterize a critical point, we will need the Jacobian matrix that f generates:

$$J = \begin{bmatrix} \frac{\partial f_0}{\partial x_0} & \frac{\partial f_0}{\partial x_1} \\ \frac{\partial f_1}{\partial x_0} & \frac{\partial f_1}{\partial x_1} \end{bmatrix} = \begin{bmatrix} 4 - 2x_1 & -2x_0 \\ x_1 & 2 + x_0 - 4x_1 \end{bmatrix} \tag{8.17}$$

Since the Jacobian describes how a function behaves in the vicinity of a point, we can use it to characterize the critical points. In Chapter 7, we used the derivative to tell us whether a point was a minimum or maximum of a function. For the Jacobian, we use the eigenvalues of J in much the same way to talk about the type and stability of critical points.

First, let's find the Jacobian at each of the critical points:

$$J\big|_{x=c_0} = \begin{bmatrix} 4 & 0 \\ 0 & 2 \end{bmatrix}, \quad J\big|_{x=c_1} = \begin{bmatrix} 2 & 0 \\ 1 & -2 \end{bmatrix}, \quad J\big|_{x=c_2} = \begin{bmatrix} 0 & -4 \\ 2 & -4 \end{bmatrix}$$

We can use NumPy to get the eigenvalues of the Jacobians:

```
>>> import numpy as np
>>> np.linalg.eig([[4,0],[0,2]])[0]
array([4.,  2.])
>>> np.linalg.eig([[2,0],[1,-2]])[0]
array([-2.,  2.])
>>> np.linalg.eig([[0,-4],[2,-4]])[0]
array([-2.+2.j, -2.-2.j])
```

We encountered np.linalg.eig in Chapter 6. The eigenvalues are the first values that eig returns, hence the [0] subscript to the function call.

For critical points of a system of autonomous differential equations, the eigenvalues indicate the points' type and stability. If both eigenvalues are real and have the same sign, the critical point is a node. If the eigenvalues are less than zero, the node is stable; otherwise, it is unstable. You can think of a stable node as a pit; if you're near it, you'll fall into it. An unstable node is like a hill; if you move away from the top, the critical point, you'll fall off. The first critical point, c_0, has positive, real eigenvalues; therefore, it represents an unstable node.

If the eigenvalues of the Jacobian are real but of opposite signs, the critical point is a saddle point. We discussed saddle points in Chapter 7. A saddle point is ultimately unstable, but in two dimensions, there's a direction where you can "fall into" the saddle and a direction where you can "fall off" the saddle. Some researchers believe most minima found when training deep neural networks are really saddle points of the loss function. We see that critical point c_1 is a saddle point, since the eigenvalues are real with opposite signs.

Finally, the eigenvalues of c_2 are complex. Complex eigenvalues indicate a spiral (also called a focus). If the real part of the eigenvalues is less than zero, the spiral is stable; otherwise, it is unstable. As the eigenvalues are complex conjugates of each other, the signs of the real parts must be the same; one can't be positive while the other is negative. For c_2, the real parts are negative, so c_2 indicates a stable spiral.

Newton's Method

I presented the critical points of Equation 8.15 by fiat. The system is easy enough that we can solve for the critical points algebraically, but that might not generally be the case. One classic method for finding the roots of a function (the places where it returns zero) is known as *Newton's method*. This is an iterative method using the first derivative and an initial guess to zero in on the root. Let's look at the method in one dimension and then extend it to two. We'll see that moving to two or more dimensions requires the use of the Jacobian.

Let's use Newton's method to find the square root of 2. To do that, we need an equation such that $f(\sqrt{2}) = 0$. A moment's thought gives us one: $f(x) = 2 - x^2$. Clearly, when $x = \sqrt{2}, f(x) = 0$.

The governing equation for Newton's method in one dimension is

$$x_{n+1} = x_n - \frac{f(x_n)}{f'(x_n)} \tag{8.18}$$

where x_0 is some initial guess at the solution.

We substitute x_0 for x_n on the right-hand side of Equation 8.18 to find x_1. We then repeat using x_1 on the right-hand side to get x_2, and so on until we see little change in x_n. At that point, if our initial guess is reasonable, we have the value we're looking for. Newton's method converges quickly, so for typical examples, we only need a handful of iterations. Of course, we have powerful computers at our fingertips, so we'll use them instead of working by hand. The Python code we need is in Listing 8-1.

```
import numpy as np
def f(x):
    return 2.0 - x*x
def d(x):
    return -2.0*x

x = 1.0
for i in range(5):
    x = x - f(x)/d(x)
    print("%2d: %0.16f" % (i+1,x))
print("NumPy says sqrt(2) = %0.16f for a deviation of %0.16f" %
    (np.sqrt(2), np.abs(np.sqrt(2)-x)))
```

Listing 8-1: Finding $\sqrt{2}$ via Newton's method

Listing 8-1 defines two functions. The first, f(x), returns the function value for a given x. The second, d(x), returns the derivative at x. If $f(x) = 2 - x^2$, then $f'(x) = -2x$.

Our initial guess is $x = 1.0$. We iterate Equation 8.18 five times, printing the current estimate of the square root of 2 each time. Finally, we use NumPy to calculate the *true* value and see how far we are from it.

Running Listing 8-1 produces

```
1: 1.5000000000000000
2: 1.4166666666666667
3: 1.4142156862745099
4: 1.4142135623746899
5: 1.4142135623730951

NumPy says sqrt(2) = 1.4142135623730951 for a
deviation of 0.0000000000000000
```

which is impressive; we get $\sqrt{2}$ to 16 decimals in only five iterations.

To extend Newton's method to vector-valued functions of vectors, like Equation 8.15, we replace the reciprocal of the derivative with the inverse of the Jacobian. Why the inverse? Recall, for a diagonal matrix, the inverse is the reciprocal of the diagonal elements. If we view the scalar derivative as a 1 × 1 matrix, then the reciprocal and inverse are the same. Equation 8.18 is already using the inverse of the Jacobian, albeit one for a 1 × 1 matrix. Therefore, we'll iterate

$$\boldsymbol{x}_{n+1} = \boldsymbol{x}_n - \boldsymbol{J}^{-1}\Big|_{\boldsymbol{x}=\boldsymbol{x}_n} f(\boldsymbol{x}_n) \tag{8.19}$$

for a suitable initial value, $\boldsymbol{x}_0$, and the inverse of the Jacobian evaluated at $\boldsymbol{x}_n$. Let's use Newton's method to find the critical points of Equation 8.15.

Before we can write some Python code, we need the inverse of the Jacobian, Equation 8.17. The inverse of a 2×2 matrix,

$$A = \begin{bmatrix} a & b \\ c & d \end{bmatrix}$$

is

$$A^{-1} = \frac{1}{\det(A)} \begin{bmatrix} d & -b \\ -c & a \end{bmatrix}$$

assuming the determinant is not zero. The determinant of A is $ad - bc$. Therefore, the inverse of Equation 8.17 is

$$J^{-1} = \frac{1}{(4 - 2x_1)(2 + x_0 - 4x_1) + 2x_0 x_1} \begin{bmatrix} 2 + x_0 - 4x_1 & 2x_0 \\ -x_1 & 4 - 2x_1 \end{bmatrix}$$

Now we can write our code. The result is Listing 8-2.

```
import numpy as np

def f(x):
    x0,x1 = x[0,0],x[1,0]
    return np.array([[4*x0-2*x0*x1],[2*x1+x0*x1-2*x1**2]])

def JI(x):
    x0,x1 = x[0,0],x[1,0]
    d = (4-2*x1)*(2-x0-4*x1)+2*x0*x1
    return (1/d)*np.array([[2-x0-4*x1,2*x0],[-x1,4-2*x0]])

x0 = float(input("x0: "))
x1 = float(input("x1: "))
❶ x = np.array([[x0],[x1]])

N = 20
for i in range(N):
    ❷ x = x - JI(x) @ f(x)
    if (i > (N-10)):
        print("%4d: (%0.8f, %0.8f)" % (i, x[0,0],x[1,0]))
```

Listing 8-2: Newton's method in 2D

Listing 8-2 echoes Listing 8-1 for the 1D case. We have f(x) to calculate the function value for a given input vector and JI(x) to give us the value of the inverse Jacobian at *x*. Notice that f(x) returns a column vector and JI(x) returns a 2×2 matrix.

The code first asks the user for initial guesses, x0 and x1. These are formed into the initial vector, x. Note that we explicitly form x as a column vector ❶.

The implementation of Equation 8.19 comes next ❷. The inverse Jacobian is a 2×2 matrix that we multiply on the right by the function value, a 2×1 column vector, using NumPy's matrix multiplication operator, @. The result is a 2×1 column vector subtracted from the current value of x, itself a 2×1 column vector. If the loop is within 10 iterations of completion, the current value is printed at the console.

Does Listing 8-2 work? Let's run it and see if we can find initial guesses leading to each of the critical points (Equation 8.16). For an initial guess of $x_0 = \begin{bmatrix} -1 \\ 2 \end{bmatrix}$, we get

```
11: (0.00004807, -1.07511237)
12: (0.00001107, -0.61452262)
13: (0.00000188, -0.27403667)
14: (0.00000019, -0.07568702)
15: (0.00000001, -0.00755378)
16: (0.00000000, -0.00008442)
17: (0.00000000, -0.00000001)
18: (0.00000000, -0.00000000)
19: (0.00000000, -0.00000000)
```

which is the first critical point of Equation 8.15. To find the two remaining critical points, we need to pick our initial guesses with some care. Some guesses explode, while many lead back to the zero vector. However, some trial and error gives us

$$\begin{bmatrix} 1 \\ 1 \end{bmatrix} \rightarrow \begin{bmatrix} 0 \\ 1 \end{bmatrix} \quad \text{and} \quad \begin{bmatrix} 1.6 \\ 1.8 \end{bmatrix} \rightarrow \begin{bmatrix} 2 \\ 2 \end{bmatrix}$$

showing that Newton's method can find the critical points of Equation 8.15.

We started this section with a system of differential equations that we interpreted as a vector-valued function. We then used the Jacobian to characterize the critical points of that system. Next, we used the Jacobian a second time to locate the system's critical points via Newton's method. We could do this because the Jacobian is the generalization of the gradient to vector-valued functions, and the gradient itself is a generalization of the first derivative of a scalar function. As mentioned above, we'll see Jacobians again when we discuss backpropagation in Chapter 10.

Concerning Hessians

If the Jacobian matrix is like the first derivative of a function of one variable, then the *Hessian matrix* is like the second derivative. In this case, we're restricted to scalar fields, functions returning a scalar value for a vector input.

Let's start with the definition and go from there. For the function $f(x)$, the Hessian is defined as

$$H_f = \begin{bmatrix} \dfrac{\partial^2 f}{\partial x_0^2} & \dfrac{\partial^2 f}{\partial x_0 \partial x_1} & \cdots & \dfrac{\partial^2 f}{\partial x_0 \partial x_{n-1}} \\[2ex] \dfrac{\partial^2 f}{\partial x_1 \partial x_0} & \dfrac{\partial^2 f}{\partial x_1^2} & \cdots & \dfrac{\partial^2 f}{\partial x_1 \partial x_{n-1}} \\[1ex] \vdots & \vdots & & \vdots \\[1ex] \dfrac{\partial^2 f}{\partial x_{n-1} \partial x_0} & \dfrac{\partial^2 f}{\partial x_{n-1} \partial x_1} & \cdots & \dfrac{\partial^2 f}{\partial x_{n-1}^2} \end{bmatrix} \tag{8.20}$$

where $x = \begin{bmatrix} x_0 & x_1 & \cdots & x_{n-1} \end{bmatrix}^\top$.

Looking at Equation 8.20 tells us that the Hessian is a square matrix. Moreover, it's a symmetric matrix implying $H = H^\top$.

The Hessian is the Jacobian of the gradient of a scalar field:

$$H_f = J(\nabla f)$$

Let's see this with an example. Consider this function:

$$f(x) = 2x_0^2 + x_0 x_2 + 3x_1 x_2 - x_1^2$$

If we use the definition of the Hessian in Equation 8.20 directly, we see that $\partial^2 f / \partial x_0^2 = 4$ because $\partial f / \partial x_0 = 4x_0 + x_2$. Similar calculations give us the rest of the Hessian matrix:

$$\begin{bmatrix} 4 & 0 & 1 \\ 0 & -2 & 3 \\ 1 & 3 & 0 \end{bmatrix}$$

In this case, the Hessian is constant, not a function of x, because the highest power of a variable in $f(x)$ is 2.

The gradient of $f(x)$, using our column vector definition, is

$$\nabla f = \begin{bmatrix} \dfrac{\partial f}{\partial x_0} \\[1.5ex] \dfrac{\partial f}{\partial x_1} \\[1.5ex] \dfrac{\partial f}{\partial x_2} \end{bmatrix} = \begin{bmatrix} 4x_0 + x_2 \\ 3x_2 - 2x_1 \\ x_0 + 3x_1 \end{bmatrix}$$

with the Jacobian of the gradient giving the following, which is identical to the matrix we found by direct use of Equation 8.20.

$$J(\nabla f) = \begin{bmatrix} \dfrac{\partial f_0}{\partial x_0} & \dfrac{\partial f_0}{\partial x_1} & \dfrac{\partial f_0}{\partial x_2} \\[1.5ex] \dfrac{\partial f_1}{\partial x_0} & \dfrac{\partial f_1}{\partial x_1} & \dfrac{\partial f_1}{\partial x_2} \\[1.5ex] \dfrac{\partial f_2}{\partial x_0} & \dfrac{\partial f_2}{\partial x_1} & \dfrac{\partial f_2}{\partial x_2} \end{bmatrix} = \begin{bmatrix} 4 & 0 & 1 \\ 0 & -2 & 3 \\ 1 & 3 & 0 \end{bmatrix}$$

Minima and Maxima

We saw in Chapter 7 that we could use the second derivative to test whether critical points of a function were minima ($f'' > 0$) or maxima ($f'' < 0$). We'll see in the next section how we can use critical points in optimization problems. For now, let's use the Hessian to find critical points by considering its eigenvalues. We'll continue with the example above. The Hessian matrix is 3×3, meaning there are three (or fewer) eigenvalues. Again, we'll save time and use NumPy to tell us what they are:

```
>>> np.linalg.eig([[4,0,1],[0,-2,3],[1,3,0]])[0]
array([ 4.34211128,  1.86236874, -4.20448002])
```

Two of the three eigenvalues are positive, and one is negative. If all three were positive, the critical point would be a minimum. Likewise, if all three were negative, the critical point would be a maximum. Notice that the minimum/maximum label is the opposite of the sign, just like the single-variable case. However, if at least one eigenvalue is positive and another negative, which is the case with our example, the critical point is a saddle point.

It seems natural to ask whether the Hessian of a vector-valued function, $f(x)$, exists. After all, we can calculate the Jacobian of such a function; we did so above to show that the Hessian is the Jacobian of the gradient.

It is possible to extend the Hessian to a vector-valued function. However, the result is no longer a matrix, but an order-3 tensor. To see this is so, consider the definition of a vector-valued function:

$$f(x) = \begin{bmatrix} f_0(x) \\ f_1(x) \\ \vdots \\ f_{m-1}(x) \end{bmatrix}$$

We can think of a vector-valued function, a vector field, as a vector of scalar functions of a vector. We could calculate the Hessian of each of the m functions in f to get a vector of matrices,

$$H_f = \begin{bmatrix} H_{f_0} \\ H_{f_1} \\ \vdots \\ H_{f_{m-1}} \end{bmatrix}$$

but a vector of matrices is a 3D object. Think of an RGB image: a 3D array made up of three 2D images, one each for the red, green, and blue channels. Therefore, while possible to define and calculate, the Hessian of a vector-valued function is beyond our current scope.

Optimization

In deep learning, you're most likely to see the Hessian in reference to optimization. Training a neural network is, to a first approximation, an optimization problem—the goal is to find the weights and biases leading to a minimum in the loss function landscape.

In Chapter 7, we saw that the gradient provides information on how to move toward a minimum. An optimization algorithm, like gradient descent, the subject of Chapter 11, uses the gradient as a guide. As the gradient is a first derivative of the loss function, algorithms based solely on the gradient are known as *first-order optimization methods*.

The Hessian provides information beyond the gradient. As a second derivative, the Hessian contains information about how the loss landscape's gradient is changing, that is, its curvature. An analogy from physics might help here. A particle's motion in one dimension is described by some function of time, $x(t)$. The first derivative, the velocity, is $dx/dt = v(t)$. The velocity tells us how quickly the position is changing in time. However, the velocity might change with time, so its derivative, $dv/dt = a(t)$, is the acceleration. And, if the velocity is the first derivative of the position, then the acceleration is the second, $d^2x/dt^2 = a(t)$. Similarly, the second derivative of the loss function, the Hessian, provides information on how the gradient is changing. Optimization algorithms using the Hessian, or an approximation of it, are known as *second-order optimization methods*.

Let's start with an example in one dimension. We have a function, $f(x)$, and we're currently at some x_0. We want to move from this position to a new position, x_1, closer to a minimum of $f(x)$. A first-order algorithm will use the gradient, here the derivative, as a guide, since we know moving in the direction opposite to the derivative will move us toward a lower function value. Therefore, for some step size, call it η (eta), we can write

$$x_1 = x_0 - \eta f'(x_0)$$

This will move us from x_0 toward x_1, which is closer to the minimum of $f(x)$, assuming the minimum exists.

The equation above makes sense, so why think about a second-order method? The second-order method comes into play when we move from $f(x)$ to $f(\boldsymbol{x})$. Now we have a gradient, not just a derivative, and the landscape of $f(\boldsymbol{x})$ around some point can be more complex. The general form of gradient descent is

$$\boldsymbol{x}_1 = \boldsymbol{x}_0 - \eta \nabla f(\boldsymbol{x}_0)$$

but the information in the Hessian can be of assistance. To see how, we first need to introduce the idea of a *Taylor series expansion*, a way of approximating an arbitrary function as the sum of a series of terms. We use Taylor series frequently in physics and engineering to simplify complex functions in the vicinity of a specific point. We also often use them to calculate values of *transcendental functions* (functions that can't be written as a finite set of basic algebra operations). For example, it's likely that when you use cos(x) in a programming language, the result is generated by a Taylor series expansion with a sufficient number of terms to get the cosine to 32- or 64-bit floating-point precision:

$$\cos(x) = \sum_{k=0}^{\infty} (-1)^k \frac{x^{2k}}{(2k)!} \approx 1 - \frac{x^2}{2!} + \frac{x^4}{4!} - \frac{x^6}{6!} + \frac{x^8}{8!}$$

In general, to approximate a function, $f(x)$, in the vicinity of a point, $x = a$, the Taylor series expansion is

$$f(x) = \sum_{k=0}^{\infty} \frac{f^{(k)}(a)}{k!} (x - a)^k$$

where $f^{(k)}(a)$ is the kth derivative of $f(x)$ evaluated at point a.

A linear approximation of $f(x)$ around $x = a$ is

$$f(x) \approx f(a) + f'(a)(x - a)$$

while a quadratic approximation of $f(x)$ becomes

$$f(x) \approx f(a) + f'(a)(x - a) + \frac{1}{2} f''(a)(x - a)^2 \tag{8.21}$$

where we see the linear approximation using the first derivative and the quadratic using the first and second derivatives of $f(x)$. A first-order optimization algorithm uses the linear approximation, while a second-order one uses the quadratic approximation.

Moving from a scalar function of a scalar, $f(x)$, to a scalar function of a vector, $f(\boldsymbol{x})$, changes the first derivative to a gradient and the second derivative to the Hessian matrix,

$$f(\boldsymbol{x}) \approx f(\boldsymbol{a}) + (\boldsymbol{x} - \boldsymbol{a})^\top \nabla f(\boldsymbol{a}) + \frac{1}{2} (\boldsymbol{x} - \boldsymbol{a})^\top \boldsymbol{H}_f(\boldsymbol{a})(\boldsymbol{x} - \boldsymbol{a})$$

with $\boldsymbol{H}_f(\boldsymbol{a})$ the Hessian matrix for $f(\boldsymbol{x})$ evaluated at the point $\boldsymbol{a}$. The products' order changes to make the dimensions work out properly, as we now have vectors and a matrix to deal with.

For example, if $\boldsymbol{x}$ has n elements, then $f(\boldsymbol{a})$ is a scalar; the gradient at $\boldsymbol{a}$ is an n-element column vector multiplying $(\boldsymbol{x} - \boldsymbol{a})^\top$, an n-element row vector, producing a scalar; and the last term is $1 \times n$ times $n \times n$ times $n \times 1$, leading to $1 \times n$ times $n \times 1$, which is also a scalar.

To use the Taylor series expansions for optimization, to find the minimum of f, we can use Newton's method in much the same way used in Equation 8.18. First, we rewrite Equation 8.21 to change our viewpoint to one of a displacement (Δx) from a current position (x). Equation 8.21 then becomes

$$f(x + \Delta x) \approx f(x) + f'(x)\Delta x + \frac{1}{2} f''(x)(\Delta x)^2 \tag{8.22}$$

Equation 8.22 is a parabola in Δx, and we're using it as a stand-in for the more complex shape of f in the region of $x + \Delta x$. To find the minimum of Equation 8.22, we take the derivative and set it to zero, then solve for Δx. The derivative gives

$$\frac{d}{d(\Delta x)} f(x + \Delta x) \approx f'(x) + f''(x)\Delta x$$

which, if set to zero, leads to

$$\Delta x = -\frac{f'(x)}{f''(x)} \tag{8.23}$$

Equation 8.23 tells us the offset from a current position, x, that would lead to the minimum of $f(x)$ if $f(x)$ were actually a parabola. In reality, $f(x)$ isn't a parabola, so the Δx of Equation 8.23 isn't the actual offset to the minimum of $f(x)$. However, since the Taylor series expansion used the actual slope, $f'(x)$, and curvature, $f''(x)$, of $f(x)$ at x, the offset of Equation 8.23 is a better estimate of the actual minimum of $f(x)$ than the linear approximation, assuming there is a minimum.

If we go from x to $x + \Delta x$, there's no reason why we can't then use Equation 8.23 a second time, calling the new position x. Thinking like this leads to an equation we can iterate:

$$x_{n+1} = x_n - \frac{f'(x_n)}{f''(x_n)} \tag{8.24}$$

for x_0, some initial starting point.

We can work out all of the above for scalar functions with vector arguments, $f(\mathbf{x})$, which are the kind we most often encounter in deep learning via the loss function. Equation 8.24 becomes

$$\mathbf{x}_{n+1} = \mathbf{x}_n - \mathbf{H}_f^{-1}\Big|_{\mathbf{x}_n} \nabla f(\mathbf{x}_n)$$

where the reciprocal of the second derivative becomes the inverse of the Hessian matrix evaluated at $\mathbf{x}_n$.

Excellent! We have an algorithm we can use to rapidly find the minimum of a function like $f(\mathbf{x})$. We saw above that Newton's method converges quickly, so using it to minimize a loss function also should converge quickly, faster than gradient descent, which only considers the first derivative.

If this is the case, why do we use gradient descent to train neural networks instead of Newton's method?

There are several reasons. First, we haven't discussed issues arising from the Hessian's applicability, issues related to the Hessian being a positive definite matrix. A symmetric matrix is positive definite if all its eigenvalues are positive. Near saddle points, the Hessian might not be positive definite, which can cause the update rule to move away from the minimum. As you might expect with a simple algorithm like Newton's method, some variations try to address issues like this, but even if problems with the eigenvalues of the Hessian are addressed, the computational burden of using the Hessian for updating network parameters is what stops Newton's algorithm in its tracks.

Every time the network's weights and biases are updated, the Hessian changes, requiring it and its inverse to be calculated again. Think of the number of minibatches used during network training. For even one minibatch, there are k parameters in the network, where k is easily in the millions to even billions. The Hessian is a $k \times k$ symmetric and positive definite matrix. Inverting the Hessian typically uses Cholesky decomposition, which is more efficient than other methods but is still an $\mathcal{O}(k^3)$ algorithm. The *big-O* notation indicates that the algorithm's resource use scales as the cube of the size of the matrix in time, memory, or both. This means doubling the number of parameters in the network increases the computational time to invert the Hessian by a factor of $2^3 = 8$ while tripling the number of parameters requires some $3^3 = 27$ times as much effort, and quadrupling some $4^3 = 64$ times as much. And this is to say nothing about storing the k^2 elements of the Hessian matrix, all floating-point values.

The computation necessary to use Newton's method with even modest deep networks is staggering. Gradient-based, first-order optimization methods are about all we can use for training neural networks.

NOTE *This statement is perhaps a bit premature. Recent work in the area of* neuroevolution *has demonstrated that evolutionary algorithms can successfully train deep models. My experimentation with swarm optimization techniques and neural networks lends credence to this approach as well.*

That first-order methods work as well as they do seems, for now, to be a very happy accident.

Some Examples of Matrix Calculus Derivatives

We conclude the chapter with some examples similar to the kinds of derivatives we commonly find in deep learning.

Derivative of Element-Wise Operations

Let's begin with the derivative of element-wise operations, which includes things like adding two vectors together. Consider

$$f = a + b = \begin{bmatrix} f_0 \\ f_1 \\ \vdots \\ f_{n-1} \end{bmatrix} = \begin{bmatrix} a_0 + b_0 \\ a_1 + b_1 \\ \vdots \\ a_{n-1} + b_{n-1} \end{bmatrix}$$

which is the straightforward addition of two vectors, element by element. What does $\partial f / \partial a$, the Jacobian of f, look like? From the definition, we have

$$\frac{\partial f}{\partial a} = \begin{bmatrix} \frac{\partial f_0}{\partial a_0} & \frac{\partial f_0}{\partial a_1} & \cdots & \frac{\partial f_0}{\partial a_{n-1}} \\ \frac{\partial f_1}{\partial a_0} & \frac{\partial f_1}{\partial a_1} & \cdots & \frac{\partial f_1}{\partial a_{n-1}} \\ \vdots & \vdots & & \vdots \\ \frac{\partial f_{n-1}}{\partial a_0} & \frac{\partial f_{n-1}}{\partial a_1} & \cdots & \frac{\partial f_{n-1}}{\partial a_{n-1}} \end{bmatrix}$$

but f_0 only depends on a_0, while f_1 depends on a_1, and so on. Therefore, all derivatives $\partial f_i / \partial a_j$ for $i \neq j$ are zero. This removes all the off-diagonal elements of the matrix, leaving

$$\frac{\partial f}{\partial a} = \begin{bmatrix} \frac{\partial f_0}{\partial a_0} & 0 & \cdots & 0 \\ 0 & \frac{\partial f_1}{\partial a_1} & \cdots & 0 \\ \vdots & \vdots & & \vdots \\ 0 & 0 & \cdots & \frac{\partial f_{n-1}}{\partial a_{n-1}} \end{bmatrix} = \begin{bmatrix} 1 & 0 & \cdots & 0 \\ 0 & 1 & \cdots & 0 \\ \vdots & \vdots & & \vdots \\ 0 & 0 & \cdots & 1 \end{bmatrix} = I$$

since $\partial f_i / \partial a_i = 1$ for all i. Similarly, $\partial f / \partial b = I$ as well. Also, if we change from addition to subtraction, $\partial f / \partial a = I$, but $\partial f / \partial b = -I$.

If the operator is element-wise multiplication of a and b, $f = a \otimes b$, then we get the following, where the diag(x) notation means the n elements of vector x along the diagonal of an $n \times n$ matrix that is zero elsewhere.

$$\frac{\partial f}{\partial a} = \begin{bmatrix} \frac{\partial(a_0 b_0)}{\partial a_0} & 0 & \cdots & 0 \\ 0 & \frac{\partial(a_1 b_1)}{\partial a_1} & \cdots & 0 \\ \vdots & \vdots & & \vdots \\ 0 & 0 & \cdots & \frac{\partial(a_{n-1} b_{n-1})}{\partial a_{n-1}} \end{bmatrix} = \begin{bmatrix} b_0 & 0 & \cdots & 0 \\ 0 & b_1 & \cdots & 0 \\ \vdots & \vdots & & \vdots \\ 0 & 0 & \cdots & b_{n-1} \end{bmatrix} = \text{diag}(b)$$

$$\frac{\partial f}{\partial b} = \begin{bmatrix} \frac{\partial(a_0 b_0)}{\partial b_0} & 0 & \cdots & 0 \\ 0 & \frac{\partial(a_1 b_1)}{\partial b_1} & \cdots & 0 \\ \vdots & \vdots & & \vdots \\ 0 & 0 & \cdots & \frac{\partial(a_{n-1} b_{n-1})}{\partial b_{n-1}} \end{bmatrix} = \begin{bmatrix} a_0 & 0 & \cdots & 0 \\ 0 & a_1 & \cdots & 0 \\ \vdots & \vdots & & \vdots \\ 0 & 0 & \cdots & a_{n-1} \end{bmatrix} = \text{diag}(a)$$

Derivative of the Activation Function

Let's find the derivative of the weights and bias value for a single node of a hidden layer in a feedforward network. Recall, the inputs to the node are the outputs of the previous layer, x, multiplied term by term by the weights, w, and summed along with the bias value, b, a scalar. The result, a scalar, is passed to the activation function to produce the output value for the node. Here, we're using the *rectified linear unit (ReLU)* which returns its argument if the argument is positive. If the argument is negative, ReLU returns zero. We can write this process as

$$y = \text{ReLU}(\boldsymbol{w} \cdot \boldsymbol{x} + b) \tag{8.25}$$

In order to implement backpropagation, we need the derivatives of Equation 8.25 with respect to $\boldsymbol{w}$ and b. Let's see how to find them.

We begin by considering the pieces of Equation 8.25. For example, from Equation 8.9, we know the derivative of the dot product with respect to $\boldsymbol{w}$ is

$$\frac{\partial(\boldsymbol{x} \cdot \boldsymbol{w})}{\partial \boldsymbol{w}} = \boldsymbol{x}^\top$$

where we have taken advantage of the fact that the dot product is commutative, $\boldsymbol{w} \cdot \boldsymbol{x} = \boldsymbol{x} \cdot \boldsymbol{w}$. Also, since b does not depend on $\boldsymbol{w}$, we have

$$\frac{\partial(\boldsymbol{w} \cdot \boldsymbol{x} + b)}{\partial \boldsymbol{w}} = \frac{\partial(\boldsymbol{w} \cdot \boldsymbol{x})}{\partial \boldsymbol{w}} + \frac{\partial b}{\partial \boldsymbol{w}} = \boldsymbol{x}^\top + 0 = \boldsymbol{x}^\top \tag{8.26}$$

What about the derivative of ReLU? By definition,

$$\text{ReLU}(z) = \max(0, z) = \begin{cases} 0, & z \le 0 \\ z, & z > 0 \end{cases}$$

implying that

$$\frac{\partial}{\partial z}\text{ReLU}(z) = \begin{cases} 0, & z \le 0 \\ 1, & z > 0 \end{cases} \tag{8.27}$$

since $\partial z / \partial z = 1$.

To find the derivatives of Equation 8.25 with respect to $\boldsymbol{w}$ and b, we need the chain rule and the results above. Let's start with $\boldsymbol{w}$. The chain rule tells us how

$$\frac{\partial y}{\partial \boldsymbol{w}} = \frac{\partial y}{\partial z}\frac{\partial z}{\partial \boldsymbol{w}}$$

with $z = \boldsymbol{w} \cdot \boldsymbol{x} + b$ and $y = \text{ReLU}(z)$.

We know $\partial y / \partial z$; it's the two cases above for the ReLU, Equation 8.27. So now we have

$$\frac{\partial y}{\partial \boldsymbol{w}} = \begin{cases} 0\frac{\partial z}{\partial \boldsymbol{w}}, & z \le 0 \\ 1\frac{\partial z}{\partial \boldsymbol{w}}, & z > 0 \end{cases}$$

and we know $\partial z/\partial \boldsymbol{w} = \boldsymbol{x}^\top$; it's Equation 8.26. Therefore, our final result is

$$\frac{\partial y}{\partial \boldsymbol{w}} = \begin{cases} 0, & \boldsymbol{w} \cdot \boldsymbol{x} + b \leq 0 \\ \boldsymbol{x}^\top, & \boldsymbol{w} \cdot \boldsymbol{x} + b > 0 \end{cases}$$

where we've replaced z with $\boldsymbol{w} \cdot \boldsymbol{x} + b$.

We follow much the same procedure to find $\partial y/\partial b$, as

$$\frac{\partial y}{\partial b} = \frac{\partial y}{\partial z} \frac{\partial z}{\partial b}$$

but $\partial y/\partial z$ is 0 or 1, depending on the z's sign. Likewise, $\partial z/\partial b = 1$, which leads to

$$\frac{\partial y}{\partial b} = \begin{cases} 0, & \boldsymbol{w} \cdot \boldsymbol{x} + b \leq 0 \\ 1, & \boldsymbol{w} \cdot \boldsymbol{x} + b > 0 \end{cases}$$

Summary

In this dense chapter, we learned about matrix calculus, including working with derivatives of functions involving vectors and matrices. We worked through the definitions and discussed some identities. We then introduced the Jacobian and Hessian matrices as analogs for first and second derivatives and learned how to use them in optimization problems. Training a deep neural network is, fundamentally, an optimization problem, so the potential utility of the Jacobian and Hessian is clear, even if the latter can't be easily used for large neural networks. We ended the chapter with some examples for derivatives of expressions found in deep learning.

This concludes the general mathematics portion of the book. We'll now turn our attention to using what we've learned to understand the workings of deep neural networks. Let's begin with a discussion of how data flows through a neural network model.

9

DATA FLOW IN NEURAL NETWORKS

In this chapter, I'll present how data flows through a trained neural network. In other words, we'll look at how to go from an input vector or tensor to the output, and the form the data takes along the way. If you're already familiar with how neural networks function, great, but if not, walking through how data flows from layer to layer will help you build an understanding of the processes involved.

First, we'll look at how we represent data in two different kinds of networks. Then, we'll work through a traditional feedforward network to give ourselves a solid foundation. We'll see just how compact inference with a neural network can be in terms of code. Finally, we'll follow data through a convolutional neural network by introducing convolutional and pooling layers. The goal of this chapter isn't to present how popular toolkits pass data around. The toolkits are highly optimized pieces of software, and such low-level knowledge isn't helpful to us at this stage. Instead, the goal is to help you see how the data flows from input to output.

Representing Data

In the end, everything in deep learning is about data. We have data that we're using to create a model, which we test with more data, ultimately letting us make predictions about even more data. We'll start by looking at how we represent data in two types of neural networks: traditional neural networks and deep convolutional networks.

Traditional Neural Networks

For a *traditional neural network* or other classical machine learning models, the input is a vector of numbers, the feature vector. The training data is a collection of these vectors, each with an associated label. (We'll restrict ourselves to basic supervised learning in this chapter.) A collection of feature vectors is conveniently implemented as a matrix, where each row is a feature vector and the number of rows matches the number of samples in the dataset. As we now know, a computer conveniently represents a matrix using a 2D array. Therefore, when working with traditional neural networks or other classical models (support vector machines, random forests, and so on), we'll represent datasets as 2D arrays.

For example, the iris dataset, which we first encountered in Chapter 6, has four features in each feature vector. We represented it as a matrix:

```
>>> import numpy as np
>>> from sklearn import datasets
>>> iris = datasets.load_iris()
>>> X = iris.data[:5]
>>> X
array([[5.1, 3.5, 1.4, 0.2],
       [4.9, 3. , 1.4, 0.2],
       [4.7, 3.2, 1.3, 0.2],
       [4.6, 3.1, 1.5, 0.2],
       [5. , 3.6, 1.4, 0.2]])
>>> Y = iris.target[:5]
```

Here, we've shown the first five samples as we did in Chapter 6. The samples above are all for class 0 (*I. setosa*). To pass this knowledge to the model, we need a matching vector of class labels; X[i] returns the feature vector for sample i, and Y[i] returns the class label. The class label is usually an integer and counts up from zero for each class in the dataset. Some toolkits prefer one-hot-encoded class labels, but we can easily create them from the more standard integer labels.

Therefore, a traditional dataset uses matrices between layers to hold weights, with the input and output of each layer a vector. This is straightforward enough. What about a more modern, deep network?

Deep Convolutional Networks

Deep networks might use feature vectors, especially if the model implements 1D convolutions, but more often than not, the entire point of using a deep network is to allow *convolutional layers* to take advantage of spatial relationships in the data. Usually, this means the inputs are images, which we represent using 2D arrays. But the input doesn't always need to be an image. The model is blissfully unaware of *what* the input represents; only the model designer knows, and they decide the architecture based on that knowledge. For simplicity, we'll assume the inputs are images, since we're already aware of how computers work with images, at least at a high level.

A black-and-white image, or one with shades of gray, known as a grayscale image, uses a single number to represent each pixel's intensity. Therefore, a grayscale image consists of a single matrix represented in the computer as a 2D array. However, most of the images we see on our computers are color images, not grayscale. Most software represents a pixel's color by three numbers: the amount of red, the amount of green, and the amount of blue. This is the origin of the *RGB* label given to color images on a computer. There are many other ways of representing colors, but RGB is by far the most common. The blending of these primary colors allows computers to display millions of colors. If each pixel needs three numbers, then a color image isn't a single 2D array, but three 2D arrays, one for each color.

For example, in Chapter 4, we loaded a color image from sklearn. Let's look at it again to see how it's arranged in memory:

```
>>> from sklearn.datasets import load_sample_image
>>> china = load_sample_image('china.jpg')
>>> china.shape
(427, 640, 3)
```

The image is returned as a NumPy array. Asking for the shape of the array returns a tuple: (427, 640, 3). The array has three dimensions. The first is the height of the image, 427 pixels. The second is the width of the image, 640 pixels. The third is the number of *bands* or *channels*, here three because it's an RGB image. The first channel is the red component of the color of each pixel, the second the green, and the last the blue. We can look at each channel as a grayscale image if we want:

```
>>> from PIL import Image
>>> Image.fromarray(china).show()
>>> Image.fromarray(china[:,:,0]).show()
>>> Image.fromarray(china[:,:,1]).show()
>>> Image.fromarray(china[:,:,2]).show()
```

PIL refers to Pillow, Python's library for working with images. If you don't already have it installed, this will install it for you:

```
pip3 install pillow
```

Each image looks similar, but if you place them side by side, you'll notice differences. See Figure 9-1. The net effect of each per-channel image creates the actual color displayed. Replace china[:,:,0] with just china to see the full color image.

Figure 9-1: The red (left), green (middle), and blue (right) china image channels

Inputs to deep networks are often multidimensional. If the input's a color image, we need to use a 3D tensor to contain the image. We're not quite done, however. Each input sample to the model is a 3D tensor, but we seldom work with a single sample at a time. When training a deep network, we use *minibatches*, sets of samples processed together to get an average loss. This implies yet another dimension to the input tensor, one that lets us specify *which* member of the minibatch we want. Therefore, the input is a 4D tensor: $N \times H \times W \times C$, with N being the number of samples in the minibatch, H the height of each image in the minibatch, W the width of each image, and C the number of channels. We'll sometimes write this in tuple form as (N, H, W, C).

Let's take a look at some actual data meant for a deep network. The data is the CIFAR-10 dataset. It's a widely used benchmark dataset and is available here: *https://www.cs.toronto.edu/~kriz/cifar.html*. You don't need to download the raw dataset, however. We've included NumPy versions with the code for this book. As mentioned above, we need two arrays: one for the images and the other for the associated labels. You'll find them in the *cifar10_test_images.npy* and *cifar10_test_labels.npy* files, respectively. Let's take a look:

```
>>> images = np.load("cifar10_test_images.npy")
>>> labels = np.load("cifar10_test_labels.npy")
>>> images.shape
(10000, 32, 32, 3)
>>> labels.shape
(10000,)
```

Notice that the images array has four dimensions. The first is the number of images in the array ($N = 10,000$). The second and third tell us that the images are 32×32 pixels. The last tells us that there are three channels, implying the dataset consists of color images. Note that, in general, the number of channels might refer to any collection of data grouped that way—it need not be an actual image. The `labels` vector has 10,000 elements as well. These are the class labels, of which there are 10 classes, a mix of animals and vehicles. For example,

---

```
>>> labels[123]
2
>>> Image.fromarray(images[123]).show()
```

This indicates that image 123 is of class 2 (bird) and that the label is correct; the image displayed should be that of a bird. Recall that, in NumPy, asking for a single index returns the entire subarray, so `images[123]` is equivalent to `images[123,:,:,:]`. The `fromarray` method of the `Image` class converts a NumPy array to an image so `show` can display it.

Working with minibatches means we pass a subset of the entire dataset through the model. If our model uses minibatches of 24, then the input to the deep network, if using CIFAR-10, is a (24, 32, 32, 3) array: 24 images, each of which has 32 rows, 32 columns, and 3 channels. We'll see below that the idea of channels is not restricted to the input to a deep network; it also applies to the shape of the data passed between layers.

We'll return to data for deep networks shortly. But for now, let's switch gears to the more straightforward topic of dataflow in a traditional, feedforward neural network.

Data Flow in Traditional Neural Networks

As we indicated above, in a traditional neural network, the weights between layers are stored as matrices. If layer i has n nodes and layer $i - 1$ has m outputs, then the weight matrix between the two layers, W_i, is an $n \times m$ matrix. When this matrix is multiplied on the right by the $m \times 1$ column vector of outputs from layer $i - 1$, the result is an $n \times 1$ output representing the input to the n nodes for layer i. Specifically, we calculate

$$a_i = \sigma(W_i a_{i-1} + b_i)$$

where a_{i-1}, the $m \times 1$ vector of outputs from layer $i - 1$, multiplies W_i to produce an $n \times 1$ column vector. We add b_i, the bias values for layer i, to this vector and apply the activation function, σ, to every element of the resulting vector, $W_i a_{i-1} + b_i$, to produce a_i, the activations for layer i. We feed the activations to layer $i + 1$ as the output of layer i. By using matrices and vectors, the rules of matrix multiplication automatically calculate all the necessary products without explicit loops in the code.

Let's see an example with a simple neural network. We'll generate a random dataset with two features and then split this dataset into train and test groups. We'll use `sklearn` to train a simple feedforward neural network on the training set. The network has a single hidden layer with five nodes and uses a rectified linear activation function (ReLU). We'll then test the trained network to see how well it learned and, most importantly, look at the actual weight matrices and bias vectors.

To build the dataset, we'll select a set of points in 2D space that are clustered but slightly overlapping. We want the network to have to learn something that isn't completely trivial. Here is the code:

```
from sklearn.neural_network import MLPClassifier

np.random.seed(8675309)
❶ x0 = np.random.random(50)-0.3
  y0 = np.random.random(50)+0.3
  x1 = np.random.random(50)+0.3
  y1 = np.random.random(50)-0.3
  x = np.zeros((100,2))
  x[:50,0] = x0; x[:50,1] = y0
  x[50:,0] = x1; x[50:,1] = y1
❷ y = np.array([0]*50+[1]*50)

❸ idx = np.argsort(np.random.random(100))
  x = x[idx]; y = y[idx]
  x_train = x[:75]; x_test = x[75:]
  y_train = y[:75]; y_test = y[75:]
```

We need the MLPClassifier class from sklearn, so we load it first. We then define a 2D dataset, x, consisting of two clouds of 50 points each. The points are randomly distributed (x0, y0 and x1, y1) but centered at (0.2, 0.8) and (0.8, 0.2), respectively ❶. Note, we set the NumPy random number seed to a fixed value, so each run produces the same set of numbers we'll see below. Feel free to remove this line and experiment with how well the network trains for various generations of the dataset.

We know the first 50 points in x are from what we'll call class 0, and the next 50 points are class 1, so we define a label vector, y ❷. Finally, we randomize the order of the points in x ❸, being careful to alter the labels in the same way, and we split them into a training set (x_train) and labels (y_train) and a test set (x_test) and labels (y_test). We keep 75 percent of the data for training and leave the remaining 25 percent for testing.

Figure 9-2 shows a plot of the full dataset, with each feature on one of the axes. The circles correspond to class 0 instances and the squares to class 1 instances. There is clear overlap between the two classes.

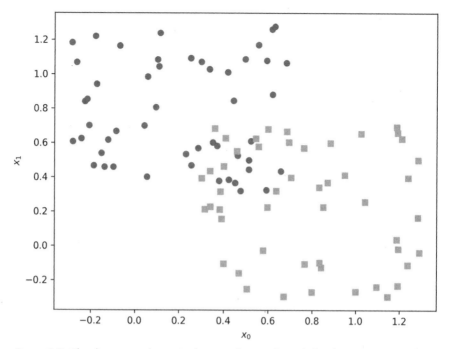

Figure 9-2: The dataset used to train the neural network, with the class 0 instances shown as circles and the class 1 instances as squares

We're now ready to train the model. The sklearn toolkit makes it easy for us, if we use the defaults:

```
❶ clf = MLPClassifier(hidden_layer_sizes=(5,))
  clf.fit(x_train, y_train)

❷ score = clf.score(x_test, y_test)
  print("Model accuracy on test set: %0.4f" % score)

❸ W0 = clf.coefs_[0].T
  b0 = clf.intercepts_[0].reshape((5,1))
  W1 = clf.coefs_[1].T
  b1 = clf.intercepts_[1]
```

Training involves creating an instance of the model class ❶. Notice that by using the defaults, which include using a ReLU activation function, we only need to specify the number of nodes in the hidden layers. We want one hidden layer with five nodes, so we pass in the tuple (5,). Training is a single call to the fit function passing in the training data, x_train, and the associated labels, y_train. When complete, we test the model by computing the accuracy (score) on the test set (x_test, y_test) and display the result.

Neural networks are initialized randomly, but because we fixed the NumPy random number seed when we generated the dataset, and because sklearn uses the NumPy random number generator as well, the outcome of training the network should be the same for each run of the code. The model has an accuracy of 92 percent on the test data ❷. This is convenient for us but concerning as well—so many toolkits use NumPy under the hood that interactions due to fixing the random number seed are probable, usually undesirable, and perhaps challenging to detect.

We're now finally ready to get the weight matrices and bias vectors from the trained network ❸. Because sklearn uses np.dot for matrix multiplication, we take the transpose of the weight matrices, W0 and W1, to get them in a form that's easier to follow mathematically. We'll see precisely why this is necessary below. Likewise, b0, the bias vector for the hidden layer, is a 1D NumPy array, so we change it to a column vector. The output layer bias, b1, is a scalar, as there is only one output for this network, the value we pass to the sigmoid function to get the probability of class 1 membership.

Let's walk through the network for the first test sample. To save space, we'll only show the first three digits of the numeric values, but our calculations will use full precision. The input to the network is

$$x = \begin{bmatrix} 0.252 \\ 1.092 \end{bmatrix}$$

We want the network to give us an output leading to the likelihood of this input belonging to class 1.

To get the output of the hidden layer, we multiply x by the weight matrix, $\boldsymbol{W_0}$, add the bias vector, $\boldsymbol{b_0}$, and pass that result through the ReLU:

$$
\begin{aligned}
\boldsymbol{a_0} &= \text{ReLU}(\boldsymbol{W_0}x + \boldsymbol{b}) \\
&= \text{ReLU}\left(\begin{bmatrix} 0.111 & 1.018 \\ 0.419 & -0.547 \\ 0.137 & 0.615 \\ -0.427 & -0.225 \\ -0.786 & -0.472 \end{bmatrix} \begin{bmatrix} 0.252 \\ 1.092 \end{bmatrix} + \begin{bmatrix} -0.055 \\ 0.238 \\ -0.280 \\ -0.313 \\ 0.901 \end{bmatrix} \right) \\
&= \text{ReLU}\left(\begin{bmatrix} 1.084 \\ -0.254 \\ 0.427 \\ -0.667 \\ 0.187 \end{bmatrix} \right) \\
&= \begin{bmatrix} 1.084 \\ 0. \\ 0.427 \\ 0. \\ 0.187 \end{bmatrix}
\end{aligned}
$$

The hidden layer to output transition uses the same form, with a_0 in place of x, but here, there is no ReLU applied:

$$a_1 = W_1 a_0 + b_1$$

$$= \begin{bmatrix} -0.383 & 1.227 & -0.938 & 0.329 & -0.638 \end{bmatrix} \begin{bmatrix} 1.084 \\ 0. \\ 0.427 \\ 0. \\ 0.187 \end{bmatrix} + 0.340$$

$$= -0.59575099$$

To get the final output probability, we use a_1, a scalar value, as the argument to the *sigmoid function*, also called the *logistic function*:

$$\text{sigmoid}(a_1) = \frac{1}{1 + e^{-a_1}} = \frac{1}{1 + e^{0.59575099}} = 0.3553164$$

This means the network has assigned a 35.5 percent likelihood of the input value being a member of class 1. The usual threshold for class assignment for a binary model is 50 percent, so the network would assign x to class 0. A peek at y_test[0] tells us the network is correct in this case: x is from class 0.

Data Flow in Convolutional Neural Networks

We saw above how data flow through a traditional neural network was straightforward matrix-vector math. To track data flow through a *convolutional neural network (CNN)*, we need to learn first what the convolution operation is and how it works. Specifically, we'll learn how to pass data through convolutional and pooling layers to a fully connected layer at the top of the model. This sequence accounts for many CNN architectures, at least at a conceptual level.

Convolution

Convolution involves two functions and the *sliding* of one over the other. If the functions are $f(x)$ and $g(x)$, convolution is defined as

$$(f * g)(t) \equiv \int_{-\infty}^{\infty} f(\tau) g(t - \tau) d\tau \tag{9.1}$$

Fortunately for us, we're working in a discrete domain and more often than not with 2D inputs, so the integral is not actually used, though $*$ is still a useful notation for the operation.

The net effect of Equation 9.1 is to slide $g(x)$ over $f(x)$ for different shifts. Let's clarify using a 1D, discrete example.

Convolution in One Dimension

Figure 9-3 shows a plot on the bottom and two sets of numbers labeled f and g on the top.

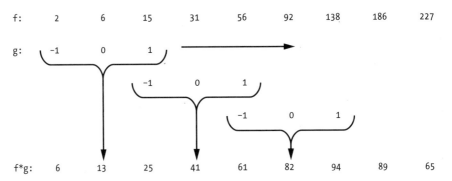

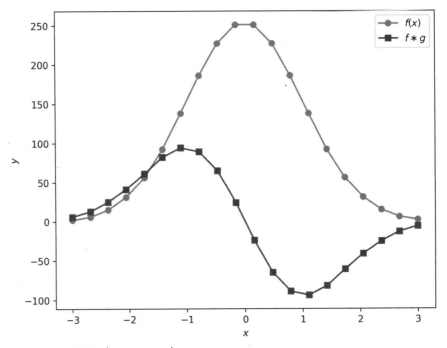

Figure 9-3: A 1D, discrete convolution

Let's start with the numbers shown at the top of Figure 9-3. The first row lists the discrete values of f. Below that is g, a three-element linear ramp. Convolution aligns g with the left edge of f as shown. We multiply corresponding elements between the two arrays,

$$[2, 6, 15] \times [-1, 0, 1] = [-2, 0, 15]$$

and then sum the resulting values,

$$-2 + 0 + 15 = 13$$

to produce the value that goes in the indicated element of the output, $f * g$. To complete the convolution, g slides one element to the right, and the process repeats. Note that in Figure 9-3, we're showing every other alignment of f and g for clarity, so it'll appear as though g is sliding two elements to the right. In general, we refer to g as a *kernel*, the set of values that slide over the input, f.

The plot on the bottom of Figure 9-3 is $f(x) = \lfloor 255 \exp(-0.5x^2) \rfloor$ for x in $[-3, 3]$ at the points marked with circles. The floor operation makes the output an integer to simplify the discussion below.

The square points in Figure 9-3 are the output of the convolution of $f(x)$ with $g(x) = [-1, 0, 1]$.

The f and $f * g$ points in Figure 9-3 are generated via

```
x = np.linspace(-3,3,20)
f = (255*np.exp(-0.5*x**2)).astype("int32")
g = np.array([-1,0,1])
fp= np.convolve(f,g[::-1], mode='same')
```

This code requires some explanation.

First, we have x, a vector spanning $[-3, 3]$ in 20 steps; this vector generates f ($f(x)$ above). We want f to be of integer type, which is what astype does for us. Next, we define g, the small linear ramp. As we'll see, the convolution operation slides g over the elements of f to produce the output.

The convolution operation comes next. As convolution is commonly used, NumPy supplies a 1D convolution function, np.convolve. The first argument is f, and the second is g. I'll explain shortly why we added [::-1] to g to reverse it. I'll also explain the meaning of mode='same'. The output of the convolution is stored in fp.

The first position shown in the top part of Figure 9-3 fills in the 13 in the output. Where does the 6 to the left of the 13 come from? Convolution has issues at the edges of f, where the kernel does not entirely cover the input. For a three-element kernel, there will be one edge element on each end of f. Kernels typically have an odd number of values, so there is a clear middle element. If g had five elements, there would be two elements on each end of f that g wouldn't cover.

Convolution functions need to make a choice about these edge cases. One option would be to return only the valid portion of the convolution, to ignore the edge cases. If we had used this approach, called *valid convolution*, the output, yp, would start with element 13 and be two less in length than the input, y.

Another approach is to fill in missing values in f with zero. This is known as *zero padding*, and we typically use it to make the output of a convolution operation the same size as the input.

Using mode='same' with np.convolve selects zero padding. This explains the 6 to the left of the 13. It's what we get when adding a zero before the 2 in *f* and applying the kernel:

$$[0, 2, 6] \times [-1, 0, 1] = [0, 0, 6], \quad 0 + 0 + 6 = 6$$

If we wanted only the valid output values, we would have used mode='valid' instead.

The call to np.convolve above didn't use g. We passed g[::-1] instead, the reverse of g. We did this to make np.convolve act like the convolutions used in deep neural networks. From a mathematical and signal processing perspective, convolution uses the reverse of the kernel. The np.convolve function, therefore, reverses the kernel, meaning we need to reverse it beforehand to get the effect we want. To be technical, if we perform the operation we've called *convolution* without flipping the kernel, we're actually performing *cross-correlation*. This issue seldom comes up in deep learning because we *learn* the kernel elements during training—we don't assign them ahead of time. With that in mind, any flipping of the kernel by the toolkit process implementing the convolution operation won't affect the outcome, because the learned kernel values were learned with that flip in place. We'll assume going forward that there is no flip and, when necessary, we'll flip the kernels we give to NumPy and SciPy functions. Additionally, we'll continue to use the term *convolution* in this no-flip-of-the-kernel deep learning sense.

In general, convolution with discrete inputs involves placing the kernel over the input starting on the left, multiplying matching elements, summing, and putting the result in the output at the point where the center of the kernel matches. The kernel then slides one element to the right, and the process repeats. We can extend the discrete convolution operation to two dimensions. Most modern deep CNNs use 2D kernels, though it's possible to use 1D and 3D kernels as well.

Convolution in Two Dimensions

Convolution with a 2D kernel requires a 2D array. Images are 2D arrays of values, and convolution is a common image processing operation. For example, let's load an image, the face of the raccoon we saw in Chapter 3, and alter it with a 2D convolution. Consider the following:

```
from scipy.signal import convolve2d
from scipy.misc import face

img = face(True)
img = img[:512,(img.shape[1]-612):(img.shape[1]-100)]

k = np.array([[1,0,0],[0,-8,0],[0,0,3]])
c = convolve2d(img, k, mode='same')
```

Here, we're using the SciPy `convolve2d` function from the `signal` module. First, we load the raccoon image and subset it to a 512×512-pixel image of the raccoon's face (`img`). Next, we define a 3 × 3 kernel, `k`. Lastly, we convolve the kernel, as it is, with the face image, storing the result in `c`. The `mode='same'` keyword zero pads the image to handle the edge cases.

The code above leads to

```
img[:8,:8]:
    [[ 88  97 112 127 116  97  84  84]
     [ 62  70 100 131 126  88  52  51]
     [ 41  46  87 127 146 116  78  56]
     [ 42  45  76 107 145 137 112  76]
     [ 58  59  69  79 111 106  90  68]
     [ 74  73  68  60  72  74  72  67]
     [ 92  87  75  63  57  74  91  93]
     [105  97  85  74  60  79 102 110]]

k:
    [[ 1  0  0]
     [ 0 -8  0]
     [ 0  0  3]]

c[1:8,1:8]:
    [[-209 -382 -566 -511 -278  -69 -101]
     [-106 -379 -571 -638 -438 -284 -241]
     [-168 -391 -484 -673 -568 -480 -318]
     [-278 -357 -332 -493 -341 -242 -143]
     [-335 -304 -216 -265 -168 -165 -184]
     [-389 -307 -240 -197 -274 -396 -427]
     [-404 -331 -289 -215 -368 -476 -488]]
```

Here, we're showing the upper 8 × 8 corner of the image and the valid portion of the convolution. Recall, the valid portion is the part where the kernel completely covers the input array.

For the kernel and the image, the first valid convolution output is −209. Mathematically, the first step is element-wise multiplication with the kernel,

$$\begin{bmatrix} 88 & 97 & 112 \\ 62 & 70 & 100 \\ 41 & 46 & 87 \end{bmatrix} \cdot \begin{bmatrix} 3 & 0 & 0 \\ 0 & -8 & 0 \\ 0 & 0 & 1 \end{bmatrix} = \begin{bmatrix} 264 & 0 & 0 \\ 0 & -560 & 0 \\ 0 & 0 & 87 \end{bmatrix}$$

followed by a summation,

$$264 + 0 + 0 + 0 + (-560) + 0 + 0 + 0 + 87 = -209$$

Notice how the kernel used wasn't `k` as we defined it. Instead, `convolve2d` flipped the kernel top to bottom and then left to right before it was applied. The remainder of `c` flows from moving the kernel one position to the right and repeating the multiplication and addition. At the end of a row, the kernel moves down one position and back to the left, until the entire image has been processed. Deep learning toolkits refer to this motion as the *stride*, and it need not be one position or equal in the horizontal and vertical directions.

Figure 9-4 shows the effect of the convolution.

Figure 9-4: The original raccoon face image (left) and the convolution result (right)

To make the image, c was shifted up, so the minimum value was zero, and then divided by the maximum to map to $[0, 1]$. Finally, the output was multiplied by 255 and displayed as a grayscale image. The original face image is on the left. The convolved image is on the right. Convolution of the image with the kernel has altered the image, emphasizing some features while suppressing others.

Convolving kernels with images isn't merely an exercise to help us understand the convolution operation. It's of profound importance in the training of CNNs. Conceptually, a CNN consists of two main parts: a set of convolution and other layers taught to learn a new representation of the input, and a top-level classifier taught to use the new representation to classify the inputs. It's the joint learning of the new representation and the classifier that makes CNNs so powerful. The key to learning a new representation of the input is the set of learned convolution kernels. How the kernels alter the input as data flows through the CNN creates the new representation. Training with gradient descent and backpropagation teaches the network which kernels to create.

We're now in a position to follow data through a CNN's convolutional layers. Let's take a look.

Convolutional Layers

Above, we discussed how deep networks pass tensors from layer to layer and how the tensor usually has four dimensions, $N \times H \times W \times C$. To follow data through a convolutional layer, we'll ignore N, knowing that what we discuss is applied to each sample in the tensor. This leaves us with inputs to the convolutional layer that are $H \times W \times C$, a 3D tensor.

The output of a convolutional layer is another 3D tensor. The height and width of the output depend on the convolution kernels' particulars and how we decide to handle the edges. We'll use valid convolution for the examples here, meaning we'll discard parts of the input that the kernel doesn't wholly cover. If the kernel is 3×3, the output will be two less in height and width, one less for each edge. A 5×5 kernel loses four in height and width, two less for each edge.

The convolutional layer uses sets of *filters* to accomplish its goal. A filter is a stack of kernels. We need one filter for each of the desired output channels. The number of kernels in the stack of each filter matches the number of channels in the input. So, if the input has M channels, and we want N output channels using $K \times K$ kernels, we need N filters, each of which is a stack of M $K \times K$ kernels.

Additionally, we have a bias value for each of the N filters. We'll see below how the bias is used, but we now know how many parameters we need to learn to implement a convolutional layer with M input channels, $K \times K$ kernels, and N outputs. It's $K \times K \times M \times N$ for N filters with $K \times K \times M$ parameters each, plus N bias terms—one per filter.

Let's make all of this concrete. We have a convolutional layer. The input to the layer is an $(H, W, C) = (5, 5, 2)$ tensor, meaning a height and width of five and two channels. We'll use a 3×3 kernel with valid convolution, so the output in height and width is 3×3 from the 5×5 input. We get to select the number of output channels. Let's use three. Therefore, we need to use convolution and kernels to map a $(5,5,2)$ input to a $(3,3,3)$ output. From what we discussed above, we know we need three filters, and each filter has $3 \times 3 \times 2$ parameters, plus a bias term.

Our input stack is

$$
0: \begin{bmatrix}
2 & 1 & -1 & 0 & 3 \\
3 & 0 & 2 & 2 & 0 \\
-1 & 2 & -1 & 3 & 1 \\
3 & 1 & 3 & -1 & -2 \\
2 & 1 & 0 & 0 & 3
\end{bmatrix}
$$

$$
1: \begin{bmatrix}
1 & 1 & 0 & -2 & 3 \\
2 & 1 & -1 & -1 & 3 \\
-2 & -3 & -3 & 0 & 1 \\
0 & 0 & 1 & 0 & 1 \\
-1 & -1 & 0 & -1 & 2
\end{bmatrix}
$$

We've split the third dimension to show the two input channels, each 5×5.

The three filters are

$$
\begin{array}{ccc}
f_0 & f_1 & f_2
\end{array}
$$

$$
0: \quad
\begin{bmatrix}
-1 & 0 & -1 \\
1 & 1 & 1 \\
2 & 2 & 2
\end{bmatrix}
\quad
\begin{bmatrix}
2 & 2 & 1 \\
-1 & 2 & 1 \\
-2 & -2 & 0
\end{bmatrix}
\quad
\begin{bmatrix}
3 & 2 & 1 \\
-1 & 0 & -1 \\
-2 & -2 & 3
\end{bmatrix}
$$

$$
1: \quad
\begin{bmatrix}
1 & 1 & 1 \\
0 & 0 & 0 \\
-1 & -1 & -1
\end{bmatrix}
\quad
\begin{bmatrix}
1 & 0 & 1 \\
1 & 1 & 1 \\
0 & 0 & 1
\end{bmatrix}
\quad
\begin{bmatrix}
1 & 0 & 0 \\
0 & 1 & 0 \\
1 & 0 & -1
\end{bmatrix}
$$

Again, we've separated the third dimension. Notice how each filter has two 3×3 kernels, one for each channel of the $5 \times 5 \times 2$ input.

Let's work through applying the first filter, f_0. We need to convolve the first channel of the input with the first kernel of f_0:

$$
\begin{bmatrix}
2 & 1 & -1 & 0 & 3 \\
3 & 0 & 2 & 2 & 0 \\
-1 & 2 & -1 & 3 & 1 \\
3 & 1 & 3 & -1 & -2 \\
2 & 1 & 0 & 0 & 3
\end{bmatrix}
\times
\begin{bmatrix}
-1 & 0 & -1 \\
1 & 1 & 1 \\
2 & 2 & 2
\end{bmatrix}
=
\begin{bmatrix}
4 & 11 & 8 \\
9 & 8 & 1 \\
15 & 0 & 6
\end{bmatrix}
$$

Then, we need to convolve the second input channel with the second kernel of f_0:

$$
\begin{bmatrix}
1 & 1 & 0 & -2 & 3 \\
2 & 1 & -1 & -1 & 3 \\
-2 & -3 & -3 & 0 & 1 \\
0 & 0 & 1 & 0 & 1 \\
-1 & -1 & 0 & -1 & 2
\end{bmatrix}
\times
\begin{bmatrix}
1 & 1 & 1 \\
0 & 0 & 0 \\
-1 & -1 & -1
\end{bmatrix}
=
\begin{bmatrix}
10 & 5 & 4 \\
1 & -2 & -1 \\
-6 & -4 & -3
\end{bmatrix}
$$

Finally, we add the two convolution outputs along with the single bias scalar:

$$
f_0: \quad
\begin{bmatrix}
4 & 11 & 8 \\
9 & 8 & 1 \\
15 & 0 & 6
\end{bmatrix}
+
\begin{bmatrix}
10 & 5 & 4 \\
1 & -2 & -1 \\
-6 & -4 & -3
\end{bmatrix}
=
\begin{bmatrix}
14 & 16 & 11 \\
10 & 6 & 0 \\
9 & -4 & 3
\end{bmatrix}
+ 1 =
\begin{bmatrix}
15 & 17 & 12 \\
11 & 7 & 1 \\
10 & -3 & 4
\end{bmatrix}
$$

We now have the first 3×3 output.

Repeating the process above for f_1 and f_2 gives

$$
f_1: \quad
\begin{bmatrix}
4 & 4 & 6 \\
0 & -7 & 15 \\
-5 & 6 & 2
\end{bmatrix}
, \quad
f_2: \quad
\begin{bmatrix}
-1 & 2 & -9 \\
11 & -13 & -2 \\
-16 & 2 & 6
\end{bmatrix}
$$

We've completed the convolutional layer and generated the $3 \times 3 \times 3$ output.

Many toolkits make it easy to add operations in the call that sets up the convolutional layer, but, conceptually, these are layers of their own that accept the $3 \times 3 \times 3$ output as an input. For example, if requested, Keras will apply a ReLU to the output. Applying a ReLU, a nonlinearity, to the output of the convolution would give us

$$f_0 : \begin{bmatrix} 15 & 17 & 12 \\ 11 & 7 & 1 \\ 10 & 0 & 4 \end{bmatrix}, \; f_1 : \begin{bmatrix} 4 & 4 & 6 \\ 0 & 0 & 15 \\ 0 & 6 & 2 \end{bmatrix}, \; f_2 : \begin{bmatrix} 0 & 2 & 0 \\ 11 & 0 & 0 \\ 0 & 2 & 6 \end{bmatrix}$$

Note that all elements less than zero are now zero. We use a nonlinearity between convolutional layers for the same reason we use a nonlinear activation function in a traditional neural network: to keep the convolutional layers from collapsing into a single layer. Notice how the operation to generate the filter outputs is purely linear; each output element is a linear combination of input values. Adding the ReLU breaks this linearity.

One reason for the creation of convolutional layers was to reduce the number of learned parameters. For the example above, the input was $5 \times 5 \times 2 = 50$ elements. The desired output was $3 \times 3 \times 3 = 27$ elements. A fully connected layer between these would need to learn $50 \times 27 = 1{,}350$ weights, plus another 27 bias values. However, the convolutional layer learned three filters, each with $3 \times 3 \times 2$ weights, as well as three bias values, for a total of $3(3 \times 3 \times 2) + 3 = 57$ parameters. Adding the convolutional layer saved learning some 1,300 additional weights.

The output of a convolutional layer is often the input to a pooling layer. Let's consider that type of layer next.

Pooling Layers

Convolutional networks often use *pooling layers* after convolutional layers. Their use is a bit controversial, as they discard information, and the loss of information might make it harder for the network to learn spatial relationships. Pooling is generally performed in the spatial domain along the input tensor's height and width while preserving the number of channels.

The pooling operation is straightforward: you move a window over the image, usually 2×2 with a stride of two, to group values. The specific pooling operation performed on each group is either max or average. The max-pooling operation preserves the maximum value in the window and discards the rest. Average pooling takes the mean of the values in the window.

A 2×2 window with a stride of two results in a reduction of a factor of two in each spatial direction. Therefore, a $(24,24,32)$ input tensor leads to a $(12,12,32)$ output tensor. Figure 9-5 illustrates the process for maximum pooling.

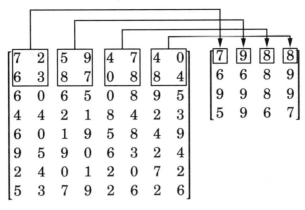

Figure 9-5: Max pooling with a 2 × 2 window and a stride of two

One channel of the input, with a height and width of eight, is on the left. The 2 × 2 window slides over the input, jumping by two, so there is no overlap of windows. The output for each 2 × 2 region of the input is the maximum value. Average pooling would instead output the mean of the four numbers. As with normal convolution, at the end of the row, the window slides down two positions, and the process repeats to change the 8 × 8 input channel to a 4 × 4 output channel.

As mentioned above, pooling without overlap in the windows loses spatial information. This has caused some in the deep learning community, most notably Geoffrey Hinton, to lament its use, as dropping spatial information distorts the relationship between objects or parts of objects in the input. For example, applying a 2 × 2 max pooling window with a stride of one instead of two to the input matrix of Figure 9-5 produces

$$\begin{bmatrix} 7 & 8 & 9 & 9 & 8 & 8 & 8 \\ 6 & 8 & 8 & 7 & 8 & 9 & 9 \\ 6 & 6 & 6 & 8 & 8 & 9 & 9 \\ 6 & 4 & 9 & 9 & 8 & 8 & 9 \\ 9 & 9 & 9 & 9 & 8 & 8 & 9 \\ 9 & 9 & 9 & 6 & 6 & 7 & 7 \\ 5 & 7 & 9 & 9 & 6 & 7 & 7 \end{bmatrix}$$

This is a 7 × 7 output, which only loses one row and column of the original 8 × 8 input. In this case, the input matrix was randomly generated, so we should expect a max-pooling operation biased toward eights and nines—there is no structure to capture. This is not usually the case in an actual CNN, of course, as it's the spatial structure inherent in the inputs we wish to utilize.

Pooling is commonly used in deep learning, especially for CNNs, so it's essential to understand what a pooling operation is doing and be aware of its potential pitfalls. Let's move on now to the output end of a CNN, typically the fully connected layers.

Fully Connected Layers

A fully connected layer in a deep network is, in terms of weights and data, identical to a regular layer in a traditional neural network. Many deep networks concerned with classification pass the output of a set of convolution and pooling layers to the first fully connected layer via a layer that flattens the tensor, essentially unraveling it into a vector. Once the output is a vector, the fully connected layer uses a weight matrix in the same way a traditional neural network does to map a vector input to a vector output.

Data Flow Through a Convolutional Neural Network

Let's put all the pieces together to see how data flows through a CNN from input to output. We'll use a simple CNN trained on the MNIST dataset, a collection of 28×28-pixel grayscale images of handwritten digits. The architecture is shown next.

$$\text{Input} \rightarrow \text{Conv(32)} \rightarrow \text{Conv(64)} \rightarrow \text{Pool} \rightarrow \text{Flatten} \rightarrow \text{Dense(128)} \rightarrow \text{Dense(10)}$$

The input is a 28×28-pixel grayscale image (one channel). The convolutional layers (conv) use 3×3 kernels and valid convolution, so their output's height and width are two less than their input. The first convolutional layer learns 32 filters while the second learns 64. We're ignoring layers that do not affect the amount of data in the network, like the ReLU layers after the convolutional layers. The max-pooling layer is assumed to use a 2×2 window with a stride of two. The first fully connected layer (dense) has 128 nodes, followed by an output layer of 10 nodes, one for each digit, 0 to 9.

The tensors passed through this network for a single input sample are

$$\underset{\text{Input}}{(28,28,1)} \rightarrow \underset{\text{Conv}}{(26,26,32)} \rightarrow \underset{\text{Conv}}{(24,24,64)} \rightarrow \underset{\text{Pool}}{(12,12,64)} \rightarrow \underset{\text{Flatten}}{9216} \rightarrow \underset{\text{Dense}}{128} \rightarrow \underset{\text{Dense}}{10}$$

The flatten layer unravels the (12,12,64) tensor to form a vector of 9,216 elements ($12 \times 12 \times 64 = 9,216$). We pass the 9,216 elements that the flatten layer outputs through the first dense layer to generate 128 output values, and the last step takes the 128-element vector and maps it to 10 output values.

Note, the values above refer to the *data* passed through the network for each input sample, one of the N samples in the minibatch. This is not the same as the number parameters (weights and biases) the network needed to learn during training.

The network shown above was trained on the MNIST digits using Keras. Figure 9-6 illustrates the action of the network for two inputs by showing, visually, the output of each layer. Specifically, it shows each layer's output for two input images, depicting a 4 and a 6.

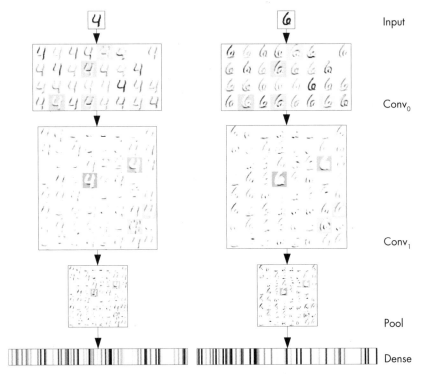

Figure 9-6: A visual representation of the output of a CNN for two sample inputs

Starting at the top, we see the two inputs. For the figure, intensities have been reversed, so darker represents higher numeric values. The input is a (28,28,1) tensor, the 1 indicating a single-channel grayscale image. Valid convolution with a 3×3 kernel returns a 26×26 output. The first convolutional layer learned 32 filters, so the output is a (26,26,32) tensor. In the figure, we show the output of each filter as an image. Zero is scaled to midlevel gray (intensity 128), more positive values are darker, and more negative values are lighter. We see differences in how the inputs have been affected by the learned filters. The single input channel means each filter in this layer is a single 3×3 kernel. Transitions between light and dark indicate edges in particular orientations.

We pass the (26,26,32) tensor through a ReLU (not shown here) and then through the second convolutional layer. The output of this layer is a (24,24,64) tensor shown as an 8×8 grid of images in the figure. We can see many parts of the input digits highlighted.

The pooling layer preserves the number of channels but reduces the spatial dimension by two. In image form, the 8×8 grid of 24×24-pixel images is now an 8×8 grid of 12×12-pixel images. The flatten operation maps the (12,12,64) tensor to a 9,216-element vector.

The output of the first dense layer is a vector of 128 numbers. For Figure 9-6, we show this as a 128-element bar code. The values run from left to right. The height of each bar is unimportant and was selected only to make the bar code easy to see. The bar code generated from the input image is the final representation that the top layer of 10 nodes uses to create the output passed through the softmax function. The highest softmax output is used to select the class label, "4" or "6."

Therefore, we can think of all the CNN layers through the first dense layer as mapping inputs to a new representation, one that makes it easy for a simple classifier to handle. Indeed, if we pass 10 examples of "4" and "6" digits through this network and display the resulting 128-node feature vectors, we get Figure 9-7, where we can easily see the difference between the digit patterns.

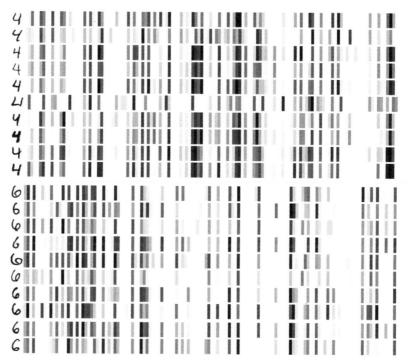

Figure 9-7: The first fully connected layer outputs for multiple "4" and "6" inputs

Of course, the entire point of writing digits as we do is to make it easy for humans to see the differences between them. While we could teach ourselves to differentiate digits using the 128-element vector images, we naturally prefer to use the written digits because of habitual use and the fact we already employ highly sophisticated hierarchical feature detectors via our brain's visual system.

The example of a CNN learning a new input representation that's more conducive to interpretation by a machine is worth bearing in mind, since what a human might use in an image as a clue to its classification is not necessarily what a network learns to use. This might explain, in part, why certain preprocessing steps, like the changes made to training samples during data augmentation, are so effective in helping the network learn to generalize, when many of those alterations seem strange to us.

Summary

The goal of this chapter was to demonstrate how neural networks manipulate data from input to output. Naturally, we couldn't cover all network types, but, in general, the principles are the same: for traditional neural networks, data is passed from layer to layer as a vector, and for deep networks, it's passed as a tensor, typically of four dimensions.

We learned how to present data to a network, either as a feature vector or a multidimensional input. We followed this by looking at how to pass data through a traditional neural network. We saw how the vectors used as input to, and output from, a layer made the implementation of a traditional neural network a straightforward exercise in matrix-vector multiplication and addition.

Next, we saw how a deep convolutional network passes data from layer to layer. We learned first about the convolution operation and then about the specifics of how convolutional and pooling layers manipulate data as tensors—a 3D tensor for each sample in the input minibatch. At the top of a CNN meant for classification are fully connected layers, which we saw act precisely as they do in a traditional neural network.

We ended the chapter by showing, visually, how input images moved through a CNN to produce an output representation, allowing the network to label the inputs correctly. We briefly discussed what this process might mean in terms of what a network picks up on during training and how that might differ from what a human naturally sees in an image.

We are now in a position to discuss backpropagation, the first of the two critical algorithms that, together with gradient descent, make training deep neural networks possible.

10

BACKPROPAGATION

Backpropagation is currently *the* core algorithm behind deep learning. Without it, we cannot train deep neural networks in a reasonable amount of time, if at all. Therefore, practitioners of deep learning need to understand what backpropagation is, what it brings to the training process, and how to implement it, at least for simple networks. For the purposes of this chapter, I'll assume you have no knowledge of backpropagation.

We'll begin the chapter by discussing what backpropagation is and what it isn't. We'll then work through the math for a trivial network. After that, we'll introduce a matrix description of backpropagation suitable for building fully connected feedforward neural networks. We'll explore the math and experiment with a NumPy-based implementation.

Deep learning toolkits like TensorFlow don't implement backpropagation the way we will in the first two sections of this chapter. Instead, they use computational graphs, which we'll discuss at a high level to conclude the chapter.

What Is Backpropagation?

In Chapter 7, we introduced the idea of the gradient of a scalar function of a vector. We worked with gradients again in Chapter 8 and saw their connection to the Jacobian matrix. Recall in that chapter, we discussed how training a neural network is essentially an optimization problem. We know training a neural network involves a loss function, a function of the network's weights and biases that tells us how well the network performs on the training set. When we do gradient descent, we'll use the gradient to decide how to move from one part of the loss landscape to another to find where the network performs best. The goal of training is to minimize the loss function over the training set.

That's the high-level picture. Now let's make it a little more concrete. Gradients apply to functions that accept vector inputs and return a scalar value. For a neural network, the vector input is the weights and biases, the parameters that define how the network performs once the architecture is fixed. Symbolically, we can write the loss function as $L(\theta)$, where θ (theta) is a vector of all the weights and biases in the network. Our goal is to move through the space that the loss function defines to find the minimum, the specific θ leading to the smallest loss, L. We do this by using the gradient of $L(\theta)$. Therefore, to train a neural network via gradient descent, we need to know how each weight and bias value contributes to the loss function; that is, we need to know $\partial L / \partial w$, for some weight (or bias) w.

Backpropagation is the algorithm that tells us what $\partial L / \partial w$ is for each weight and bias of the network. With the partial derivatives, we can apply gradient descent to improve the network's performance on the next pass of the training data.

Before we go any further, a word on terminology. You'll often hear machine learning folks use *backpropagation* as a proxy for the entire process of training a neural network. Experienced practitioners understand what they mean, but people new to machine learning are sometimes a bit confused. To be explicit, *backpropagation* is the algorithm that finds the contribution of each weight and bias value to the network's error, the $\partial L / \partial w$'s. *Gradient descent* is the algorithm that uses the $\partial L / \partial w$'s to modify the weights and biases to improve the network's performance on the training set.

Rumelhart, Hinton, and Williams introduced backpropagation in their 1986 paper "Learning Representations by Back-propagating Errors." Ultimately, backpropagation is an application of the chain rule we discussed in Chapters 7 and 8. Backpropagation begins at the network's output with the loss function. It moves *backward*, hence the name "backpropagation," to ever-lower layers of the network, propagating the error signal to find $\partial L / \partial w$ for each weight and bias. Note, practitioners frequently shorten the name to "backprop." You'll encounter that term often.

We'll work through backpropagation by example in the following two sections. For now, the primary thing to understand is that it is the first of two pieces we need to train neural networks. It provides the information required by the second piece, gradient descent, the subject of Chapter 11.

Backpropagation by Hand

Let's define a simple neural network, one that accepts two input values, has two nodes in its hidden layer, and has a single output node, as shown in Figure 10-1.

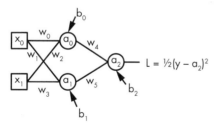

Figure 10-1: A simple neural network

Figure 10-1 shows the network with its six weights, w_0 through w_5, and three bias values, b_0, b_1, and b_2. Each value is a scalar.

We'll use sigmoid activation functions in the hidden layer,

$$\sigma(x) = \frac{1}{1 + e^{-x}}$$

and no activation function for the output node. To train the network, we'll use a squared-error loss function,

$$L = \frac{1}{2}(y - a_2)^2$$

where y is the label, zero or one, for a training example and a_2 is the output of the network for the input associated with y, namely x_0 and x_1.

Let's write the equations for a forward pass with this network, a pass that moves left to right from the input, x_0 and x_1, to the output, a_2. The equations are

$$z_0 = w_0 x_0 + w_2 x_1 + b_0$$

$$a_0 = \sigma(z_0)$$

$$z_1 = w_1 x_0 + w_3 x_1 + b_1$$

$$a_1 = \sigma(z_1)$$

$$a_2 = w_4 a_0 + w_5 a_1 + b_2 \tag{10.1}$$

Here, we've introduced intermediate values z_0 and z_1 to be the arguments to the activation functions. Notice that a_2 has no activation function. We could have used a sigmoid here as well, but as our labels are either 0 or 1, we'll learn a good output value regardless.

If we pass a single training example through the network, the output is a_2. If the label associated with the training example, $x = (x_0, x_1)$, is y, the squared-error loss is as indicated in Figure 10-1.

The argument to the loss function is a_2; y is a fixed constant. However, a_2 depends directly on w_4, w_5, b_2, and the values of a_1 and a_0, which themselves depend on $w_0, w_1, w_2, w_3, b_0, b_1, x_0$, and x_1. Therefore, thinking in terms of the weights and biases, we could write the loss function as

$$L = L(w_0, w_1, w_2, w_3, w_4, w_5, b_0, b_1, b_2; x_0, x_1, y) = L(\boldsymbol{\theta}; \boldsymbol{x}, y)$$

Here, $\boldsymbol{\theta}$ represents the weights and biases; it's considered the variable. The parts after the semicolon are constants in this case: the input vector $\boldsymbol{x} = (x_0, x_1)$ and the associated label, y.

We need the gradient of the loss function, $\nabla L(\boldsymbol{\theta}; \boldsymbol{x}, y)$. To be explicit, we need all the partial derivatives, $\partial L / \partial w_5$, $\partial L / \partial b_0$, and so on, for all weights and biases: nine partial derivatives in total.

Here's our plan of attack. First, we'll work through the math to calculate expressions for the partial derivatives of all nine values. Second, we'll write some Python code to implement the expressions so we can train the network of Figure 10-1 to classify iris flowers. We'll learn a few things during this process. Perhaps the most important is that calculating the partial derivatives by hand is, to be understated, tedious. We'll succeed, but we'll see in the following section that, thankfully, we have a far more compact way we can represent backpropagation, especially for fully connected feedforward networks. Let's get started.

Calculating the Partial Derivatives

We need expressions for all the partial derivatives of the loss function for the network in Figure 10-1. We also need an expression for the derivative of our activation function, the sigmoid. Let's begin with the sigmoid, as a clever trick writes the derivative in terms of the sigmoid itself, a value calculated during the forward pass.

The derivative of the sigmoid is shown next.

$$\sigma'(x) = \frac{d}{dx}\left(\frac{1}{1+e^{-x}}\right) = \left(\frac{-1}{(1+e^{-x})^2}\right)(-e^{-x})$$

$$= \frac{e^{-x}}{(1+e^{-x})^2}$$

$$= \left(\frac{1}{1+e^{-x}}\right)\left(\frac{e^{-x}}{1+e^{-x}}\right)$$

$$= \sigma(x)\left(\frac{e^{-x}}{1+e^{-x}}\right)$$

$$= \sigma(x)\left(\frac{1+e^{-x}-1}{1+e^{-x}}\right) \qquad (10.2)$$

$$= \sigma(x)\left(\frac{1+e^{-x}}{1+e^{-x}} - \frac{1}{1+e^{-x}}\right)$$

$$= \sigma(x)(1-\sigma(x)) \qquad (10.3)$$

The trick of Equation 10.2 is to add and subtract one in the numerator to change the form of the factor to be another copy of the sigmoid itself. So, the derivative of the sigmoid is the product of the sigmoid and one minus the sigmoid. Looking back at Equation 10.1, we see that the forward pass computes the sigmoids, the activation functions, as a_0 and a_1. Therefore, during the derivation of the backpropagation partial derivatives, we'll be able to substitute a_0 and a_1 via Equation 10.3 for the derivative of the sigmoid to avoid calculating it a second time.

Let's start with the derivatives. True to backpropagation's name, we'll work backward from the loss function and apply the chain rule to arrive at the expressions we need. The derivative of the loss function,

$$L = \frac{1}{2}(y - a_2)^2$$

is

$$\frac{\partial L}{\partial a_2} = (2)\left(\frac{1}{2}\right)(y - a_2)(-1) = a_2 - y \qquad (10.4)$$

This means that everywhere in the expressions that follow, we can replace $\partial L/\partial a_2$ with $a_2 - y$. Recall y is the label for the current training example, and we compute a_2 during the forward pass as the output of the network.

Let's now find expressions for w_5, w_4, and b_2, the parameters used to calculate a_2. The chain rule tells us

$$\frac{\partial L}{\partial w_5} = \left(\frac{\partial L}{\partial a_2}\right)\left(\frac{\partial a_2}{\partial w_5}\right) = (a_2 - y)a_1 \qquad (10.5)$$

since

$$\frac{\partial a_2}{\partial w_5} = \frac{\partial(w_4 a_0 + w_5 a_1 + b_2)}{\partial w_5} = a_1$$

We've substituted in the expression for a_2 from Equation 10.1.

Similar logic leads to expressions for w_4 and b_2:

$$\frac{\partial L}{\partial w_4} = \left(\frac{\partial L}{\partial a_2}\right)\left(\frac{\partial a_2}{\partial w_4}\right) = (a_2 - y)a_0$$

$$\frac{\partial L}{\partial b_2} = \left(\frac{\partial L}{\partial a_2}\right)\left(\frac{\partial a_2}{\partial b_2}\right) = (a_2 - y)(1) = (a_2 - y) \qquad (10.6)$$

Fantastic! We have three of the partial derivatives we need—only six more to go. Let's write the expressions for b_1, w_1, and w_3,

$$\frac{\partial L}{\partial b_1} = \left(\frac{\partial L}{\partial a_2}\right)\left(\frac{\partial a_2}{\partial a_1}\right)\left(\frac{\partial a_1}{\partial z_1}\right)\left(\frac{\partial z_1}{\partial b_1}\right) = (a_2 - y)w_5 a_1(1 - a_1)$$

$$\frac{\partial L}{\partial w_1} = \left(\frac{\partial L}{\partial a_2}\right)\left(\frac{\partial a_2}{\partial a_1}\right)\left(\frac{\partial a_1}{\partial z_1}\right)\left(\frac{\partial z_1}{\partial w_1}\right) = (a_2 - y)w_5 a_1(1 - a_1)x_0$$

$$\frac{\partial L}{\partial w_3} = \left(\frac{\partial L}{\partial a_2}\right)\left(\frac{\partial a_2}{\partial a_1}\right)\left(\frac{\partial a_1}{\partial z_1}\right)\left(\frac{\partial z_1}{\partial w_3}\right) = (a_2 - y)w_5 a_1(1 - a_1)x_1 \qquad (10.7)$$

where we use

$$\frac{\partial a_1}{\partial z_1} = \sigma'(z_1) = \sigma(z_1)(1 - \sigma(z_1)) = a_1(1 - a_1)$$

substituting a_1 for $\sigma(z_1)$ as we calculate a_1 during the forward pass.

A similar calculation gives us expressions for the final three partial derivatives:

$$\frac{\partial L}{\partial b_0} = \left(\frac{\partial L}{\partial a_2}\right)\left(\frac{\partial a_2}{\partial a_0}\right)\left(\frac{\partial a_0}{\partial z_0}\right)\left(\frac{\partial z_0}{\partial b_0}\right) = (a_2 - y)w_4 a_0(1 - a_0)$$

$$\frac{\partial L}{\partial w_0} = \left(\frac{\partial L}{\partial a_2}\right)\left(\frac{\partial a_2}{\partial a_0}\right)\left(\frac{\partial a_0}{\partial z_0}\right)\left(\frac{\partial z_0}{\partial w_0}\right) = (a_2 - y)w_4 a_0(1 - a_0)x_0$$

$$\frac{\partial L}{\partial w_2} = \left(\frac{\partial L}{\partial a_2}\right)\left(\frac{\partial a_2}{\partial a_0}\right)\left(\frac{\partial a_0}{\partial z_0}\right)\left(\frac{\partial z_0}{\partial w_2}\right) = (a_2 - y)w_4 a_0(1 - a_0)x_1 \quad (10.8)$$

Whew! That was tedious, but now we have what we need. Notice, however, that this is a very rigid process—if we change the network architecture, activation function, or loss function, we need to derive these expressions again. Let's use the expressions to classify iris flowers.

Translating into Python

The code I've presented here is in the file *nn_by_hand.py*. Take a look at it in an editor to see the overall structure. We'll start with the main function (Listing 10-1):

```
❶ epochs = 1000
   eta = 0.1

❷ xtrn, ytrn, xtst, ytst = BuildDataset()

❸ net = {}
   net["b2"] = 0.0
   net["b1"] = 0.0
   net["b0"] = 0.0
   net["w5"] = 0.0001*(np.random.random() - 0.5)
   net["w4"] = 0.0001*(np.random.random() - 0.5)
   net["w3"] = 0.0001*(np.random.random() - 0.5)
   net["w2"] = 0.0001*(np.random.random() - 0.5)
   net["w1"] = 0.0001*(np.random.random() - 0.5)
   net["w0"] = 0.0001*(np.random.random() - 0.5)

❹ tn0,fp0,fn0,tp0,pred0 = Evaluate(net, xtst, ytst)

❺ net = GradientDescent(net, xtrn, ytrn, epochs, eta)

❻ tn,fp,fn,tp,pred = Evaluate(net, xtst, ytst)

   print("Training for %d epochs, learning rate %0.5f" % (epochs, eta))
```

```
print()
print("Before training:")
print("    TN:%3d  FP:%3d" % (tn0, fp0))
print("    FN:%3d  TP:%3d" % (fn0, tp0))
print()
print("After training:")
print("    TN:%3d  FP:%3d" % (tn, fp))
print("    FN:%3d  TP:%3d" % (fn, tp))
```

Listing 10-1: The main function

First, we set the number of epochs and the learning rate, η (eta) ❶. The number of epochs is the number of passes through the training set to update the network weights and biases. The network is straightforward, and our dataset tiny, with only 70 samples, so we need many epochs for training. Gradient descent uses the learning rate to decide how to move based on the gradient values. We'll explore the learning rate more thoroughly in Chapter 11.

Next, we load the dataset ❷. We're using the same iris dataset we used in Chapter 6 and again in Chapter 9, keeping only the first two features and classes 0 and 1. See the BuildDataset function in *nn_by_hand.py*. The return values are NumPy arrays: xtrn (70×2) and xtst (30×2) for training and test data, and the associated labels in ytrn and ytst.

We need someplace to store the network weights and biases. A Python dictionary will do, so we set it up next with default values ❸. Notice that we set the bias values to zero and the weights to small random values in $[-0.00005, +0.00005]$. These seem to work well enough in this case.

The remainder of main evaluates the randomly initialized network (Evaluate ❹) on the test data, performs gradient descent to train the model (GradientDescent ❺), and evaluates the test data again to demonstrate that training worked ❻.

Listing 10-2 shows Evaluate as well as Forward, which Evaluate calls.

```
def Evaluate(net, x, y):
    out = Forward(net, x)
    tn = fp = fn = tp = 0
    pred = []
    for i in range(len(y)):
❶      c = 0 if (out[i] < 0.5) else 1
        pred.append(c)
        if (c == 0) and (y[i] == 0):
            tn += 1
        elif (c == 0) and (y[i] == 1):
            fn += 1
        elif (c == 1) and (y[i] == 0):
            fp += 1
        else:
            tp += 1
    return tn,fp,fn,tp,pred
```

```
def Forward(net, x):
    out = np.zeros(x.shape[0])
    for k in range(x.shape[0]):
❷      z0 = net["w0"]*x[k,0] + net["w2"]*x[k,1] + net["b0"]
        a0 = sigmoid(z0)
        z1 = net["w1"]*x[k,0] + net["w3"]*x[k,1] + net["b1"]
        a1 = sigmoid(z1)
        out[k] = net["w4"]*a0 + net["w5"]*a1 + net["b2"]
    return out
```

Listing 10-2: The Evaluate *function*

Let's begin with Forward, which performs a forward pass over the data in
x. After creating a place to hold the output of the network (out), each input
is run through the network using the current value of the parameters ❷.
Notice that the code is a direct implementation of Equation 10.1, with out[k]
in place of a_2. When all inputs have been processed, we return the collected
outputs to the caller.

Now let's look at Evaluate. Its arguments are a set of input features, x,
associated labels, y, and the network parameters, net. Evaluate first runs the
data through the network by calling Forward to populate out. These are the
raw, floating-point outputs from the network. To compare them with the actual labels, we apply a threshold ❶ to call outputs < 0.5 class 0 and outputs
≥ 0.5 class 1. The predicted label is appended to pred and tallied by comparing it to the actual label in y.

If the actual and predicted labels are both zero, the model has correctly
identified a *true negative* (TN), a true instance of class 0. If the network pre-
dicts class 0, but the actual label is class 1, we have a *false negative* (FN), a class
1 instance labeled class 0. Conversely, labeling a class 0 instance class 1 is
a *false positive* (FP). The only remaining option is an actual class 1 instance
labeled as class 1, a *true positive* (TP). Finally, we return the tallies and predictions to the caller.

Listing 10-3 presents GradientDescent, which Listing 10-1 calls ❺. This is
where we implement the partial derivatives calculated above.

```
def GradientDescent(net, x, y, epochs, eta):
❶ for e in range(epochs):
        dw0 = dw1 = dw2 = dw3 = dw4 = dw5 = db0 = db1 = db2 = 0.0

❷    for k in range(len(y)):
❸        z0 = net["w0"]*x[k,0] + net["w2"]*x[k,1] + net["b0"]
          a0 = sigmoid(z0)
          z1 = net["w1"]*x[k,0] + net["w3"]*x[k,1] + net["b1"]
          a1 = sigmoid(z1)
          a2 = net["w4"]*a0 + net["w5"]*a1 + net["b2"]

❹        db2 += a2 - y[k]
          dw4 += (a2 - y[k]) * a0
```

```
        dw5 += (a2 - y[k]) * a1
        db1 += (a2 - y[k]) * net["w5"] * a1 * (1 - a1)
        dw1 += (a2 - y[k]) * net["w5"] * a1 * (1 - a1) * x[k,0]
        dw3 += (a2 - y[k]) * net["w5"] * a1 * (1 - a1) * x[k,1]
        db0 += (a2 - y[k]) * net["w4"] * a0 * (1 - a0)
        dw0 += (a2 - y[k]) * net["w4"] * a0 * (1 - a0) * x[k,0]
        dw2 += (a2 - y[k]) * net["w4"] * a0 * (1 - a0) * x[k,1]

    m = len(y)
❺   net["b2"] = net["b2"] - eta * db2 / m
    net["w4"] = net["w4"] - eta * dw4 / m
    net["w5"] = net["w5"] - eta * dw5 / m
    net["b1"] = net["b1"] - eta * db1 / m
    net["w1"] = net["w1"] - eta * dw1 / m
    net["w3"] = net["w3"] - eta * dw3 / m
    net["b0"] = net["b0"] - eta * db0 / m
    net["w0"] = net["w0"] - eta * dw0 / m
    net["w2"] = net["w2"] - eta * dw2 / m

    return net
```

Listing 10-3: Using GradientDescent to train the network

The GradientDescent function contains a double loop. The outer loop ❶ is over epochs, the number of full passes through the training set. The inner loop ❷ is over the training examples, one at a time. The forward pass comes first ❸ to calculate the output, a2, and intermediate values.

The next block of code implements the backward pass using the partial derivatives, Equations 10.4 through 10.8, to move the error (loss) backward through the network ❹. We use the average loss over the training set to update the weights and biases. Therefore, we accumulate the contribution to the loss for each weight and bias value for each training example. This explains adding each new contribution to the total over the training set.

After passing each training example through the net and accumulating its contribution to the loss, we update the weights and biases ❺. The partial derivatives give us the gradient, the direction of maximal change; however, we want to minimize, so we move in the direction *opposite* to the gradient, subtracting the average of the loss due to each weight and bias from its current value.

For example,

```
net["b2"] = net["b2"] - eta * db2 / m
```

is

$$b_2 \leftarrow b_2 - \eta \left(\frac{1}{m} \sum_{i=0}^{m-1} \frac{\partial L}{\partial b_2} \Big|_{\mathbf{x}_i} \right)$$

where $\eta = 0.1$ is the learning rate and m is the number of samples in the training set. The summation is over the partial for b_2 evaluated for each input sample, x_i, the average value of which, multiplied by the learning rate, is used to adjust b_2 for the next epoch. Another name we frequently use for the learning rate is *step size*. This parameter controls how quickly the weights and biases of the network step through the loss landscape toward a minimum value.

Our implementation is complete. Let's run it to see how well it does.

Training and Testing the Model

Let's take a look at the training data. We can plot the features, one on each axis, to see how easy it might be to separate the two classes. The result is Figure 10-2, with class 0 as circles and class 1 as squares.

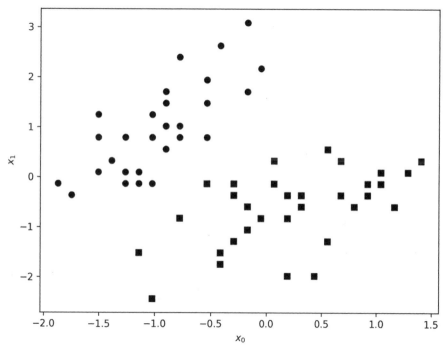

Figure 10-2: The iris training data showing class 0 (circles) and class 1 (squares)

It's straightforward to see that the two classes are quite separate from each other so that even our elementary network with two hidden neurons should be able to learn the difference between them. Compare this plot with the left side of Figure 6-2, which shows the first two features for all three iris classes. If we had included class 2 in our dataset, two features would not be enough to separate all three classes.

Run the code with

```
python3 nn_by_hand.py
```

For me, this produces

```
Training for 1000 epochs, learning rate 0.10000

Before training:
    TN: 15  FP:  0
    FN: 15  TP:  0

After training:
    TN: 14  FP:  1
    FN:  1  TP: 14
```

We're told training used 1,000 passes through the training set of 70 examples. This is the outer loop of Listing 10-3. We're then presented with two tables of numbers, characterizing the network before training and after. Let's walk through these tables to understand the story they tell.

The tables are known by several names: *contingency tables*, *2 × 2 tables*, or *confusion matrices*. The term *confusion matrix* is the most general, though it's usually reserved for multiclass classifiers. The labels count the number of true positives, true negatives, false positives, and false negatives in the test set. The test set includes 30 samples, 15 from each class. If the network is perfect, all class 0 samples will be in the TN count, and all class 1 in the TP count. Errors are FP or FN counts.

The randomly initialized network labels everything as class 0. We know this because there are 15 TN samples (those that are truly class 0) and 15 FN samples (15 class 1 samples that are labeled class 0). The overall accuracy before training is then $15/(15 + 15) = 0.5 = 50$ percent.

After training, the 1,000 passes through the outer loop of the code in Listing 10-3, the test data is almost perfectly classified, with 14 of the 15 class 0 and 14 of the 15 class 1 labels correctly assigned. The overall accuracy is now $(14 + 14)/(15 + 15) = 28/30 = 93.3$ percent—not too shabby considering our model has a single hidden layer of two nodes.

Again, this exercise's main point is to see how tedious and potentially error-prone it is to calculate derivatives by hand. The code above works with scalars; it doesn't process vectors or matrices to take advantage of any symmetry possible by using a better representation of the backpropagation algorithm. Thankfully, we can do better. Let's look again at the backpropagation algorithm for fully connected networks and see if we can use vectors and matrices to arrive at a more elegant approach.

Backpropagation for Fully Connected Networks

In this section, we'll explore the equations that allow us to pass an error term backward from the output of the network to the input. Additionally, we'll see how to use this error term to calculate the necessary partial derivatives of the weights and biases for a layer so we can implement gradient descent. With all the essential expressions on hand, we'll implement Python classes that will allow us to build and train fully connected feedforward neural networks of arbitrary depth and shape. We'll conclude by testing the classes against the MNIST dataset.

Backpropagating the Error

Let's begin with a useful observation: the layers of a fully connected neural network can be thought of as vector functions:

$$y = f(x)$$

where the input to the layer is x and the output is y. The input, x, is either the actual input to the network for a training sample or, if working with one of the hidden layers of the model, the previous layer's output. These are both vectors; each node in a layer produces a single scalar output, which, when grouped, becomes y, a vector representing the output of the layer.

The forward pass runs through the layers of the network in order, mapping x_i to y_i so that y_i becomes x_{i+1}, the input to layer $i + 1$. After all layers are processed, we use the final layer output, call it h, to calculate the loss, $L(h, y_{true})$. The loss is a measure of how wrong the network is for the input, x, that we determine by comparing it to the true label y_{true}. Note that if the model is multiclass, the output h is a vector, with one element for each possible class, and the true label is a vector of zeros, except for the index of the actual class label, which is one. This is why many toolkits, like Keras, map integer class labels to one-hot vectors.

We need to move the loss value, or the *error*, back through the network; this is the backpropagation step. To do this for a fully connected network using per-layer vectors and weight matrices, we need to first see how to run the forward pass. As we did for the network we built above, we'll separate applying the activation function from the action of a fully connected layer.

For example, for any layer with the input vector x coming from the layer below, we need to calculate an output vector, y. For a fully connected layer, the forward pass is

$$y = Wx + b$$

where W is a weight matrix, x is the input vector, and b is the bias vector.

For an activation layer, we have

$$y = \sigma(x)$$

for whatever activation function, σ, we choose. We'll stick with the sigmoid for the remainder of this chapter. Note we made the function a vector-valued function. To do this, we apply the scalar sigmoid function to each element of the input vector to produce the output vector:

$$\sigma(x) = [\sigma(x_0)\ \sigma(x_1)\ \cdots\ \sigma(x_{n-1})]^\top$$

A fully connected network consists of a series of fully connected layers followed by activation layers. Therefore, the forward pass is a chain of operations that begins with the input to the model being given to the first layer to produce an output, which is then passed to the next layer's input, and so on until all layers have been processed.

The forward pass leads to the final output and the loss. The derivative of the loss function with respect to the network output is the first error term. To pass the error term back down the model, we need to calculate how the error term changes with a change to the input of a layer using how the error changes with a change to the layer's output. Specifically, for each layer, we need to know how to calculate

$$\frac{\partial E}{\partial \boldsymbol{x}}$$

That is, we need to know how the error term changes with a change in the input to the layer given

$$\frac{\partial E}{\partial \boldsymbol{y}}$$

which is how the error term changes with a change in the output of the layer. The chain rule tells us how to do it:

$$\frac{\partial E}{\partial \boldsymbol{x}} = \frac{\partial E}{\partial \boldsymbol{y}} \frac{\partial \boldsymbol{y}}{\partial \boldsymbol{x}} \tag{10.9}$$

where $\partial E / \partial \boldsymbol{x}$ for layer i becomes $\partial E / \partial \boldsymbol{y}$ for layer $i - 1$ as we move backward through the network.

Operationally, the backpropagation algorithm becomes

1. Run a forward pass to map $\boldsymbol{x} \rightarrow \boldsymbol{y}$, layer by layer, to get the final output, $\boldsymbol{h}$.
2. Calculate the value of the derivative of the loss function using $\boldsymbol{h}$ and $\boldsymbol{y}_{true}$; this becomes $\partial E / \partial \boldsymbol{y}$ for the output layer.
3. Repeat for all earlier layers to calculate $\partial E / \partial \boldsymbol{x}$ from $\partial E / \partial \boldsymbol{y}$, causing $\partial E / \partial \boldsymbol{x}$ for layer i to become $\partial E / \partial \boldsymbol{y}$ for layer $i - 1$.

This algorithm passes the error term backward through the network. Let's work out how to get the necessary partial derivatives by layer type, beginning with the activation layer.

We will assume we know $\partial E / \partial \boldsymbol{y}$ and are looking for $\partial E / \partial \boldsymbol{x}$. The chain rule says

$$\frac{\partial E}{\partial \boldsymbol{x}} = \frac{\partial E}{\partial \boldsymbol{y}} \frac{\partial \boldsymbol{y}}{\partial \boldsymbol{x}}$$

$$= \left[\frac{\partial E}{\partial y_0} \frac{\partial y_0}{\partial x_0} \ \frac{\partial E}{\partial y_1} \frac{\partial y_1}{\partial x_1} \ \cdots \right]^\top$$

$$= \left[\frac{\partial y_0}{\partial x_0} \sigma'(x_0) \ \frac{\partial y_1}{\partial x_1} \sigma'(x_1) \ \cdots \right]^\top$$

$$= \frac{\partial E}{\partial \boldsymbol{y}} \odot \boldsymbol{\sigma}'(\boldsymbol{x}) \tag{10.10}$$

Here, we're introducing $\odot$ to represent the Hadamard product. Recall that the Hadamard product is the element-wise multiplication of two vectors or matrices. (See Chapter 5 for a refresher.)

We now know how to pass the error term through an activation layer. The only other layer we're considering is a fully connected layer. If we expand Equation 10.9, we get

$$\frac{\partial E}{\partial \boldsymbol{x}} = \frac{\partial E}{\partial \boldsymbol{y}} \frac{\partial \boldsymbol{y}}{\partial \boldsymbol{x}}$$

$$= \boldsymbol{W}^\top \frac{\partial E}{\partial \boldsymbol{y}} \tag{10.11}$$

since

$$\frac{\partial \boldsymbol{y}}{\partial \boldsymbol{x}} = \frac{\partial (\boldsymbol{W}\boldsymbol{x} + \boldsymbol{b})}{\partial \boldsymbol{x}} = \boldsymbol{W}^\top \quad \text{(denominator layout)}$$

The result is $\boldsymbol{W}^\top$, not $\boldsymbol{W}$, because the derivative of a matrix times a vector in denominator notation is the transpose of the matrix rather than the matrix itself.

Let us pause for a bit to recap and think about the form of Equations 10.10 and 10.11. These equations tell us how to pass the error term backward from layer to layer. What are the shapes of these values? For the activation layer, if the input has k-elements, then the output also has k-elements. Therefore, the relationship in Equation 10.10 should map a k-element vector to another k-element vector. The error term, $\partial E / \partial \boldsymbol{y}$, is a k-element vector, as is the derivative of the activation function, $\boldsymbol{\sigma}'(\boldsymbol{x})$. Finally, the Hadamard product between the two also outputs a k-element vector, as needed.

For the fully connected layer, we have an m-element input, x; an $n \times m$-element weight matrix, W; and an output vector, y, of n-elements. So we need to generate an m-element vector, $\partial E / \partial x$, from the n-element error term, $\partial E / \partial y$. Multiplying the transpose of the weight matrix, an $m \times n$-element matrix, by the error term does result in an m-element vector, since $m \times n$ by $n \times 1$ is $m \times 1$, an m-element column vector.

Calculating Partial Derivatives of the Weights and Biases

Equations 10.10 and 10.11 tell us how to pass the error term backward through the network. However, the point of backpropagation is to calculate how changes in the weights and biases affect the error so we can use gradient descent. Specifically, for every fully connected layer, we need expressions for

$$\frac{\partial E}{\partial W} \text{ and } \frac{\partial E}{\partial b}$$

given

$$\frac{\partial E}{\partial y} \text{ and } \frac{\partial E}{\partial x}$$

Let's start with $\partial E / \partial b$. Applying the chain rule yet again gives

$$\frac{\partial E}{\partial b} = \frac{\partial E}{\partial y} \frac{\partial y}{\partial b}$$

$$= \frac{\partial E}{\partial y} \frac{\partial (Wx + b)}{\partial b}$$

$$= \frac{\partial E}{\partial y} (0 + 1)$$

$$= \frac{\partial E}{\partial y} \tag{10.12}$$

meaning the error due to the bias term for a fully connected layer is the same as the error due to the output.

The calculation for the weight matrix is similar:

$$\frac{\partial E}{\partial W} = \frac{\partial E}{\partial y} \frac{\partial y}{\partial W}$$

$$= \frac{\partial E}{\partial y} \frac{\partial(Wx + b)}{\partial W}$$

$$= \frac{\partial E}{\partial y} \left(x^\top + 0\right)$$

$$= \frac{\partial E}{\partial y} x^\top \qquad (10.13)$$

The equation above tells us the error due to the weight matrix is a product of the output error and the input, x. The weight matrix is an $n \times m$-element matrix, as the forward pass multiplies by the m-element input vector. Therefore, the error contribution from the weights, $\partial E/\partial W$, also must be an $n \times m$ matrix. We know $\partial E/\partial y$ is an n-element column vector, and the transpose of x is an m-element row vector. The outer product of the two is an $n \times m$ matrix, as required.

Equations 10.10, 10.11, 10.12, and 10.13 apply for a single training example. This means for a specific input to the network, these equations, especially 10.12 and 10.13, tell us the contribution to the loss by the biases and weights of any layer *for that input sample*.

To implement gradient descent, we need to accumulate these errors, the $\partial E/\partial W$ and $\partial E/\partial b$ terms, over the training samples. We then use the average value of these errors to update the weights and biases at the end of every epoch or, as we'll implement it, minibatch. As gradient descent is the subject of Chapter 11, all we'll do here is outline how we use backpropagation to implement gradient descent and leave the details to that chapter and the code we'll implement next.

In general, however, to train the network, we need to do the following for each sample in the minibatch:

1. Forward pass the sample through the network to create the output. Along the way, we need to store the input to each layer, as we need it to implement backpropagation (that is, we need $x^\top$ from Equation 10.13).

2. Calculate the value of the derivative of the loss function, which for us is the mean squared error, to use as the first error term in backpropagation.

3. Run through the layers of the network in reverse order, calculating $\partial E/\partial W$ and $\partial E/\partial b$ for each fully connected layer. These values are accumulated for each sample in the minibatch (ΔW, Δb).

When the minibatch samples have been processed and the errors accumulated, it's time to take a gradient descent step. This is where the weights and biases of each layer are updated via

$$W \leftarrow W - \eta \left(\frac{1}{m} \Delta W \right)$$

$$b \leftarrow b - \eta \left(\frac{1}{m} \Delta b \right) \tag{10.14}$$

with ΔW and Δb being the accumulated errors over the minibatch and m being the size of the minibatch. Repeated gradient descent steps lead to a final set of weights and biases—a trained network.

This section is quite math-heavy. The following section translates the math into code, where we'll see that for all the math, the code, because of NumPy and object-oriented design, is quite compact and elegant. If you're fuzzy on the math, I suspect the code will go a long way toward clarifying things for you.

A Python Implementation

Our implementation is in the style of toolkits like Keras. We want the ability to create arbitrary, fully connected networks, so we'll use Python classes for each layer and store the architecture as a list of layers. Each layer maintains its weights and biases, along with the ability to do a forward pass, a backward pass, and a gradient descent step. For simplicity, we'll use sigmoid activations and the squared error loss.

We need two classes: `ActivationLayer` and `FullyConnectedLayer`. An additional `Network` class holds the pieces together and handles training. The classes are in the file *NN.py*. (The code here is modified from the original code by Omar Aflak and is used with his permission. See the GitHub link in *NN.py*. I modified the code to use minibatches and support gradient descent steps other than for every sample.)

Let's walk through each of the three classes, starting with `ActivationLayer` (see Listing 10-4). The translation of the math we've done to code form is quite elegant, in most cases a single line of NumPy.

```
class ActivationLayer:
    def forward(self, input_data):
        self.input = input_data
        return sigmoid(input_data)
    def backward(self, output_error):
        return sigmoid_prime(self.input) * output_error
    def step(self, eta):
        return
```

Listing 10-4: The ActivationLayer class

Listing 10-4 shows `ActivationLayer` and includes only three methods: forward, backward, and step. The simplest is step. It does nothing, as there's nothing for an activation layer to do during gradient descent because there are no weights or bias values.

The forward method accepts the input vector, *x*, stores it for later use, and then calculates the output vector, *y*, by applying the sigmoid activation function.

The backward method accepts $\partial E / \partial \mathbf{y}$, the output_error from the layer above. It then returns Equation 10.10 by applying the derivative of the sigmoid (sigmoid_prime) to the input set during the forward pass, multiplied element-wise by the error.

The sigmoid and sigmoid_prime helper functions are

```
def sigmoid(x):
    return 1.0 / (1.0 + np.exp(-x))
def sigmoid_prime(x):
    return sigmoid(x)*(1.0 - sigmoid(x))
```

The `FullyConnectedLayer` class is next. It's more complex than the `ActivationLayer` class, but not significantly so. See Listing 10-5.

```
class FullyConnectedLayer:
    def __init__(self, input_size, output_size):
❶       self.delta_w = np.zeros((input_size, output_size))
        self.delta_b = np.zeros((1,output_size))
        self.passes = 0
❷       self.weights = np.random.rand(input_size, output_size) - 0.5
        self.bias = np.random.rand(1, output_size) - 0.5

    def forward(self, input_data):
        self.input = input_data
❸       return np.dot(self.input, self.weights) + self.bias

    def backward(self, output_error):
        input_error = np.dot(output_error, self.weights.T)
        weights_error = np.dot(self.input.T, output_error)
        self.delta_w += np.dot(self.input.T, output_error)
        self.delta_b += output_error
        self.passes += 1
        return input_error

    def step(self, eta):
❹       self.weights -= eta * self.delta_w / self.passes
        self.bias -= eta * self.delta_b / self.passes
❺       self.delta_w = np.zeros(self.weights.shape)
        self.delta_b = np.zeros(self.bias.shape)
        self.passes = 0
```

Listing 10-5: The FullyConnectedLayer class

We tell the constructor the number of input and output nodes. The number of input nodes (input_size) specifies the number of elements in the vector coming into the layer. Likewise, output_size specifies the number of elements in the output vector.

Fully connected layers accumulate weight and bias errors over the minibatch, the $\partial E / \partial W$ terms in delta_w and the $\partial E / \partial b$ terms in delta_b ❶. Each sample processed is counted in passes.

We must initialize neural networks with random weight and bias values; therefore, the constructor sets up an initial weight matrix and bias vector using uniform random values in the range $[-0.5, 0.5]$ ❷. Notice, the bias vector is $1 \times n$, a row vector. The code flips the ordering from the equations above to match the way training samples are usually stored: a matrix in which each row is a sample and each column a feature. The computation produces the same results because scalar multiplication is commutative: $ab = ba$.

The forward method stashes the input vector for later use by backward and then calculates the output of the layer, multiplying the input by the weight matrix and adding the bias term ❸.

Only two methods remain. The backward method receives $\partial E / \partial y$ (output_error) and calculates $\partial E / \partial x$ (input_error), $\partial E / \partial W$ (weights_error), and $\partial E / \partial b$ (output_error). We add the errors to the running error total for the layer, delta_w and delta_b, for step to use.

The step method includes a gradient descent step for a fully connected layer. Unlike the empty method of ActivationLayer, the FullyConnectedLayer has plenty to do. We update the weight matrix and bias vector using the average error, as in Equation 10.14 ❹. This implements the gradient descent step over the minibatch. Finally, we reset the accumulators and counter for the next minibatch ❺.

The Network class brings everything together, as shown in Listing 10-6.

```
class Network:
    def __init__(self, verbose=True):
        self.verbose = verbose
      ❶ self.layers = []

    def add(self, layer):
      ❷ self.layers.append(layer)

    def predict(self, input_data):
        result = []
        for i in range(input_data.shape[0]):
            output = input_data[i]
            for layer in self.layers:
                output = layer.forward(output)
            result.append(output)
      ❸ return result

    def fit(self, x_train, y_train, minibatches, learning_rate, batch_size=64):
      ❹ for i in range(minibatches):
```

```
        err = 0
        idx = np.argsort(np.random.random(x_train.shape[0]))[:batch_size]
        x_batch = x_train[idx]
        y_batch = y_train[idx]
❺ for j in range(batch_size):
            output = x_batch[j]
            for layer in self.layers:
                output = layer.forward(output)

❻     err += mse(y_batch[j], output)

❼     error = mse_prime(y_batch[j], output)
            for layer in reversed(self.layers):
                error = layer.backward(error)

❽ for layer in self.layers:
            layer.step(learning_rate)
        if (self.verbose) and ((i%10) == 0):
            err /= batch_size
            print('minibatch %5d/%d   error=%0.9f' % (i, minibatches, err))
```

Listing 10-6: The Network class

The constructor for the Network class is straightforward. We set a verbose flag to toggle displaying the mean error over the minibatch during training. Successful training should show this error decreasing over time. As layers are added to the network, they are stored in layers, which the constructor initializes ❶. The add method adds layer objects to the network by appending them to layers ❷.

After the network is trained, the predict method generates output for each input sample in input_data with a forward pass through the layers of the network. Notice the pattern: the input sample is assigned to output; then the loop over layers calls the forward method of each layer, in turn passing the output of the previous layer as input to the next; and so on through the entire network. When the loop ends, output contains the output of the final layer, so it's appended to result, which is returned to the caller ❸.

Training the network is fit's job. The name matches the standard training method for sklearn. The arguments are the NumPy array of sample vectors, one per row (x_train), and their labels as one-hot vectors (y_train). The number of minibatches to train comes next. We'll discuss minibatches in a bit. We also provide the learning rate, η (eta), and an optional minibatch size, batch_size.

The fit method uses a double loop. The first is over the desired number of minibatches ❹. As we learned earlier, a minibatch is a subset of the full training set, and an epoch is one full pass through the training set. Using the entire training set is known as *batch training*, and batch training uses epochs. However, there is good reason not to do batch training, as you'll see in Chapter 11, so the concept of a *minibatch* was introduced. The typical minibatch sizes are anywhere from 16 to 128 samples at a time. Powers of two are often used to make things nice for GPU-based deep learning toolkits. For us, there's no difference between a minibatch of 64 or 63 samples in terms of performance.

We select most minibatches as sequential sets of the training data to ensure all the data is used. Here, we're being a bit lazy and instead select random subsets each time we need a minibatch. This simplifies the code and adds one more place where randomness can show its utility. That's what idx gives us, a random ordering of indices into the training set, keeping only the first batch_size worth. We then use x_batch and y_batch for the actual forward and backward passes.

The second loop is over the samples in the minibatch ❺. Samples are passed individually through the layers of the network, calling forward just as predict does. For display purposes, the actual mean squared error between the forward pass output and the sample label is accumulated for the minibatch ❻.

The backward pass begins with the output error term, the derivative of the loss function, mse_prime ❼. The pass then continues *backward* through the layers of the network, passing the previous layer's output error as input to the layer below, directly mirroring the forward pass process.

Once the loop processes all the minibatch samples ❺, it's time to take a gradient descent step based on the mean error each layer in the network accumulated over the samples ❽. The argument to step needs only the learning rate. The minibatch concludes by reporting the average error if verbose is set for every 10th minibatch.

We'll experiment with this code again in Chapter 11 as we explore gradient descent. For now, let's test it with the MNIST dataset to see how well it works.

Using the Implementation

Let's take *NN.py* for a spin. We'll use it to build a classifier for the MNIST dataset, which we first encountered in Chapter 9. The original MNIST dataset consists of 28×28-pixel grayscale images of handwritten digits with black backgrounds. It's a workhorse of the machine learning community. We'll resize the images to 14×14 pixels before turning them into vectors of 196 elements (= 14 × 14).

The dataset includes 60,000 training images and 10,000 test images. The vectors are stored in NumPy arrays; see the files in the *dataset* directory. The code to generate the dataset is in *build_dataset.py*. If you want to run the code yourself, you'll need to install Keras and OpenCV for Python first. Keras supplies the original set of images and maps the training set labels to one-hot vectors. OpenCV rescales the images from 28×28 to 14×14 pixels.

The code we need is in *mnist.py* and is shown in Listing 10-7.

```
import numpy as np
from NN import *
```

```
❶ x_train = np.load("dataset/train_images_small.npy")
  x_test  = np.load("dataset/test_images_small.npy")
  y_train = np.load("dataset/train_labels_vector.npy")
  y_test  = np.load("dataset/test_labels.npy")

❷ x_train = x_train.reshape(x_train.shape[0], 1, 14*14)
  x_train /= 255
  x_test = x_test.reshape(x_test.shape[0], 1, 14*14)
  x_test /= 255

❸ net = Network()
  net.add(FullyConnectedLayer(14*14, 100))
  net.add(ActivationLayer())
  net.add(FullyConnectedLayer(100, 50))
  net.add(ActivationLayer())
  net.add(FullyConnectedLayer(50, 10))
  net.add(ActivationLayer())

❹ net.fit(x_train, y_train, minibatches=40000, learning_rate=1.0)

❺ out = net.predict(x_test)
  cm = np.zeros((10,10), dtype="uint32")
  for i in range(len(y_test)):
      cm[y_test[i],np.argmax(out[i])] += 1

  print()
  print(np.array2string(cm))
  print()
  print("accuracy = %0.7f" % (np.diag(cm).sum() / cm.sum(),))
```

Listing 10-7: Classifying MNIST digits

Notice that we import *NN.py* right after NumPy. We load the training images, test images, and labels next ❶. The Network class expects each sample vector to be a $1 \times n$ row vector, so we reshape the training data from (60000,196) to (60000,1,196)—the same as the test data ❷. At the same time, we scale the 8-bit data from $[0, 255]$ to $[0, 1]$. This is a standard preprocessing step for image data, as doing so makes it easier for the network to learn.

Building the model comes next ❸. First, we create an instance of the Network class. Then, we add the input layer by defining a FullyConnectedLayer with 196 inputs and 100 outputs. A sigmoid activation layer follows this. We then add a second fully connected layer mapping the 100 outputs of the first layer to 50 outputs, along with an activation layer. Finally, we add a last fully connected layer mapping the 50 outputs of the previous layer to 10, the number of classes, along with adding its activation layer. This approach mimics common toolkits like Keras.

Training happens by calling fit ❹. We specify 40,000 minibatches using the default minibatch size of 64 samples. We set the learning rate to 1.0, which works well in this instance. Training takes some 17 minutes on my old Intel i5 Ubuntu system. As the model trains, the mean error over the minibatch is reported. When training is complete, we pass the 10,000 test samples through the network and calculate a 10×10 confusion matrix ❺. Recall that the rows of the confusion matrix are the true class labels, here the actual digits 0 through 9. The columns correspond to the predicted labels, the largest value of the 10 outputs for each input sample. The matrix elements are the counts of how often the true label was i, and the assigned label was j. If the model is perfect, the matrix is purely diagonal; there are no cases where the true label and model label disagree. The overall accuracy is printed last as the diagonal sum divided by the sum of the matrix, the total number of test samples.

My run of *mnist.py* produced

```
minibatch 39940/40000   error=0.003941790
minibatch 39950/40000   error=0.001214253
minibatch 39960/40000   error=0.000832551
minibatch 39970/40000   error=0.000998448
minibatch 39980/40000   error=0.002377286
minibatch 39990/40000   error=0.000850956
```

```
[[ 965    0    1    1    1    5    2    3    2    0]
 [   0 1121    3    2    0    1    3    0    5    0]
 [   6    0 1005    4    2    0    3    7    5    0]
 [   0    1    6  981    0    4    0    9    4    5]
 [   2    0    3    0  953    0    5    3    1   15]
 [   4    0    0   10    0  864    5    1    4    4]
 [   8    2    1    1    3    4  936    0    3    0]
 [   2    7   19    2    1    0    0  989    1    7]
 [   5    0    4    5    3    5    7    3  939    3]
 [   5    5    2   10    8    2    1    3    6  967]]
```

```
accuracy = 0.9720000
```

The confusion matrix is strongly diagonal, and the overall accuracy is 97.2 percent. This isn't too bad of a result for a simple toolkit like *NN.py* and a fully connected feedforward network. The largest error that the network made was confusing sevens for twos 19 times (element [7,2] of the confusion matrix). The next closest error was confusing fours for nines 15 times (element [4,9]). Both of these errors make sense: sevens and twos often look similar, as do fours and nines.

We started this chapter with a network we created that included two inputs, two nodes in the hidden layer, and an output. The file *iris.py* implements the same model by adapting the dataset to what Network expects. We won't walk through the code, but do run it. When I do, I get slightly better performance on the test set: 14 out of 15 correct for class 0 and 15 out of 15 for class 1.

Sadly, the backpropagation methods detailed here and in the previous section are not ultimately flexible enough for deep learning. Modern toolkits don't use these approaches. Let's explore what deep learning toolkits do when it comes to backpropagation.

Computational Graphs

In computer science, a *graph* is a collection of nodes (vertices) and edges connecting them. We've been using graphs all along to represent neural networks. In this section, we'll use graphs to represent expressions instead.

Consider this simple expression:

$$y = mx + b$$

To evaluate this expression, we follow agreed-upon rules regarding operator precedence. Following the rules implies a sequence of primitive operations that we can represent as a graph, as shown in Figure 10-3.

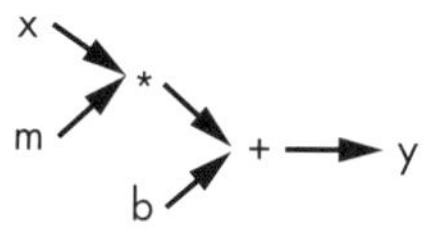

Figure 10-3: A computational graph implementing y = mx + b

Data flows through the graph of Figure 10-3 along the arrows, from left to right. Data originates in *sources*, here *x*, *m*, and *b*, and flows through *operators*, * and +, to the output, *y*.

Figure 10-3 is a *computational graph*—a graph specifying how to evaluate an expression. Compilers for languages like C generate computational graphs in some form to translate high-level expressions into sequences of machine language instructions. For the expression above, first the *x* and *m* values are multiplied, and the resulting output of the multiplication operation is passed to an addition operation, along with *b*, to produce the final output, *y*.

We can represent expressions, including those representing complex deep neural networks, as computational graphs. We represented fully connected feedforward models this way, as data flowing from the input, *x*, through the hidden layers to the output, the loss function.

Computational graphs are how deep learning toolkits like TensorFlow and PyTorch manage the structure of a model and implement backpropagation. Unlike the rigid calculations earlier in the chapter, a computational graph is generic and capable of representing all the architectures used in deep learning.

As you peruse the deep learning literature and begin to work with specific toolkits, you will run across two different approaches to using computational graphs. The first generates the graph dynamically when data is available. PyTorch uses this method, called *symbol-to-number*. TensorFlow uses the second method, *symbol-to-symbol*, to build a static computational graph ahead of time. Both approaches implement graphs, and both can automatically calculate the derivatives needed for backpropagation.

TensorFlow generates the derivatives it needs for backpropagation in much the same way we did in the previous section. Like addition, each operation knows how to create the derivative of its outputs with respect to its inputs. That, along with the chain rule, is all that's needed to implement backpropagation. Exactly how the graph is traversed depends on the *graph evaluation engine* and the specific model architecture, but the graph is traversed as needed for both the forward and backward passes. Note that because the computational graph breaks expressions into smaller operations, each of which knows how to process gradients during the backward step (as we did above for `ActivationLayer` and `FullyConnectedLayer`), it's possible to use custom functions in layers without working through the derivatives. The graph engine does it for you, as long as you use primitive operations the engine already supports.

Let's walk through the forward and backward passes of a computational graph. This example comes from the 2015 paper "TensorFlow: Large-Scale Machine Learning on Heterogeneous Distributed Systems" (*https://arxiv.org/pdf/1603.04467.pdf*).

A hidden layer in a fully connected model is expressed as

$$y = \sigma(Wx + b)$$

for weight matrix W, bias vector b, input x, and output y.

Figure 10-4 shows the same equation as a computational graph.

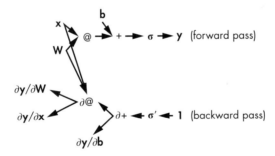

Figure 10-4: The computational graphs representing the forward and backward passes through one layer of a feedforward neural network

Figure 10-4 presents two versions. The top of the figure shows the forward pass, where data flows from x, W, and b to produce the output. Notice how the arrows lead left to right.

Note the sources are tensors, here either vectors or matrices. The outputs of operations are also tensors. The tensors flow through the graph, hence the name *TensorFlow*. Figure 10-4 represents matrix multiplication as @, the NumPy matrix multiplication operator. The activation function is σ.

For the backward pass, the sequence of derivatives begins with $\partial y/\partial y = 1$ and flows back through the graph from operator output to inputs. If there is more than one input, there is more than one output derivative. In practice, the graph evaluation engine processes the proper set of operators in the proper order. Each operator has its needed input derivatives available when it's that operator's turn to be processed.

Figure 10-4 uses ∂ before an operator to indicate the derivatives the operator generates. For example, the addition operator $(\partial+)$ produces two outputs because there are two inputs, Wx and b. The same is true for matrix multiplication $(\partial@)$. The derivative of the activation function is shown as σ'.

Notice that arrows run from W and x in the forward pass to the derivative of the matrix multiplication operator in the backward pass. Both W and x are necessary to calculate $\partial y/\partial W$ and $\partial y/\partial x$—see Equation 10.13 and Equation 10.11, respectively. There is no arrow from b to the matrix multiplication operator because $\partial y/\partial b$ does not depend on b—see Equation 10.12. If a layer were below what is shown in Figure 10-4, the $\partial y/\partial x$ output from the matrix multiplication operator would become the input for the backward pass through that layer, and so on.

The power of computational graphs makes modern deep learning toolkits highly general and supports almost any network type and architecture, without burdening the user with detailed and highly tedious gradient calculations. As you continue to explore deep learning, do appreciate what the toolkits make possible with only a few lines of code.

Summary

This chapter introduced backpropagation, one of the two pieces needed to make deep learning practical. First, we worked through calculating the necessary derivatives by hand for a tiny network and saw how laborious a process it was. However, we were able to train the tiny network successfully.

Next, we used our matrix calculus knowledge from Chapter 8 to find the equations for multilayer fully connected networks and created a simple toolkit in the same vein as toolkits like Keras. With the toolkit, we successfully trained a model to high accuracy using the MNIST dataset. While effective and general in terms of the number of hidden layers and their sizes, the toolkit was restricted to fully connected models.

We ended the chapter with a cursory look at how modern deep learning toolkits like TensorFlow implement models and automate backpropagation. The computational graph enables arbitrary combinations of primitive operations, each of which can pass gradients backward as necessary, thereby allowing the complex model architectures we find in deep learning.

The second half of training a deep model is gradient descent, which puts the gradients calculated by backpropagation to work. Let's now turn our attention that way.

11

GRADIENT DESCENT

In this final chapter, we'll slow down a bit and consider gradient descent afresh. We'll begin by reviewing the idea of gradient descent using illustrations, discussing what it is and how it works. Next, we'll explore the meaning of *stochastic* in *stochastic gradient descent*. Gradient descent is a simple algorithm that invites tweaking, so after we explore stochastic gradient descent, we'll consider a useful and commonly used tweak: momentum. We'll conclude the chapter by discussing more advanced, adaptive gradient descent algorithms, specifically RMSprop, Adagrad, and Adam.

This is a math book, but gradient descent is very much applied math, so we'll learn by experimentation. The equations are straightforward, and the math we saw in previous chapters is relevant as background. Therefore, consider this chapter an opportunity to apply what we've learned so far.

The Basic Idea

We've encountered gradient descent several times already. We know the form of the basic gradient descent update equations from Equation 10.14:

$$W \leftarrow W - \eta \, \Delta W$$

$$b \leftarrow b - \eta \, \Delta b \qquad (11.1)$$

Here, ΔW and Δb are errors based on the partial derivatives of the weights and biases, respectively; η (eta) is a step size or learning rate, a value we use to adjust how we move.

Equation 11.1 isn't specific to machine learning. We can use the same form to implement gradient descent on arbitrary functions. Let's discuss gradient descent using 1D and 2D examples to lay a foundation for how it operates. We'll use an unmodified form of gradient descent known as *vanilla gradient descent*.

Gradient Descent in One Dimension

Let's begin with a scalar function of x:

$$f(x) = 6x^2 - 12x + 3 \qquad (11.2)$$

Equation 11.2 is a parabola facing upward. Therefore, it has a minimum. Let's find the minimum analytically by setting the derivative to zero and solving for x:

$$\frac{d}{dx}\left(6x^2 - 12x + 3\right) = 12x - 12 = 0$$

$$12x = 12$$

$$x = 1$$

The minimum of the parabola is at $x = 1$. Now, let's instead use gradient descent to find the minimum of Equation 11.2. How should we begin?

First, we need to write the proper update equation, the form of Equation 11.1 that applies in this case. We need the gradient, which for a 1D function is simply the derivative, $f'(x) = 12x - 12$. With the derivative, gradient descent becomes

$$x \leftarrow x - \eta f'(x) = x - \eta \left(12x - 12\right) \qquad (11.3)$$

Notice that we subtract $\eta(12x - 12)$. This is why the algorithm is called gradient *descent*. Recall that the gradient points in the direction of maximum change in the function's value. We're interested in minimizing the function, not maximizing it, so we move in the direction opposite to the gradient toward smaller function values; therefore, we subtract.

Equation 11.3 is one gradient descent step. It moves from an initial position, x, to a new position based on the value of the slope at the current position. Again η, the learning rate, governs how far we move.

Now that we have the equation, let's implement gradient descent. We'll plot Equation 11.2, pick a starting position, say $x = -0.9$, and iterate Equation 11.3, plotting the function value at each new position of x. If we do this, we should see a series of points on the function that move ever closer to the minimum position at $x = 1$. Let's write some code.

First, we implement Equation 11.2 and its derivative:

```
def f(x):
    return 6*x**2 - 12*x + 3
def d(x):
    return 12*x - 12
```

Next, we plot the function, and then we iterate Equation 11.3, plotting the new pair, $(x, f(x))$, each time:

```
import numpy as np
import matplotlib.pylab as plt

❶ x = np.linspace(-1,3,1000)
  plt.plot(x,f(x))

❷ x = -0.9
  eta = 0.03
❸ for i in range(15):
      plt.plot(x, f(x), marker='o', color='r')
❹     x = x - eta * d(x)
```

Let's walk through the code. After importing NumPy and Matplotlib, we plot Equation 11.2 ❶. Next, we set our initial x position ❷ and take 15 gradient descent steps ❸. We plot before stepping, so we see the initial x but do not plot the last step, which is fine in this case.

The final line ❹ is key. It implements Equation 11.3. We update the current x position by multiplying the derivative's value at x by $\eta = 0.03$ as the step size. The code above is in the file *gd_1d.py*. If we run it, we get Figure 11-1.

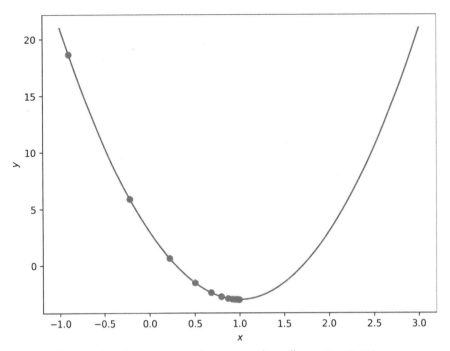

Figure 11-1: Gradient descent in one dimension with small steps (η = 0.03)

Our initial position, which we can think of as an initial guess at the location of the minimum, is $x = -0.9$. Clearly, this isn't the minimum. As we take gradient descent steps, we move successively closer to the minimum, as the sequence of circles moving toward it shows.

Notice two things here. First, we do get closer and closer to the minimum. After 14 steps, we are, for all intents and purposes, at the minimum: $x = 0.997648$. Second, each gradient descent step leads to smaller and smaller changes in x. The learning rate is constant at $\eta = 0.03$, so the source of the smaller updates to x must be smaller and smaller values of the derivative at each x position. This makes sense if we think about it. As we approach the minimum position, the derivative gets smaller and smaller, until it reaches zero at the minimum, so the update using the derivative gets successively smaller as well.

We selected the step size for Figure 11-1 to move smoothly toward the minimum of the parabola. What if we change the step size? Further along in *gd_1d.py*, the code repeats the steps above, starting at $x = 0.75$ and setting $\eta = 0.15$ to take steps that are five times larger than those plotted in Figure 11-1. The result is Figure 11-2.

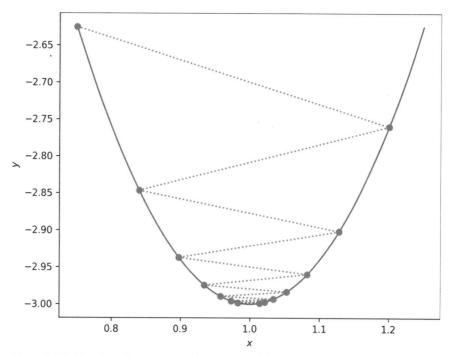

Figure 11-2: Gradient descent in one dimension with large steps ($\eta = 0.15$)

In this case, the steps overshoot the minimum. The new *x* positions oscillate, bouncing back and forth over the true minimum position. The dashed lines connect successive *x* positions. The overall search still approaches the minimum but takes longer to reach it, as the large step size makes each update to *x* tend to move past the minimum.

Small gradient descent steps move short distances along the function, whereas large steps move large distances. If the learning rate is too small, many gradient descent steps are necessary. If the learning rate is too large, the search overshoots and oscillates around the minimum position. The proper learning rate is not immediately obvious, so intuition and experience come into play when selecting it. Additionally, these examples fixed η. There's no reason why η has to be a constant. In many deep learning applications, the learning rate is not constant but evolves as training progresses, effectively making η a function of the number of gradient descent steps taken.

Gradient Descent in Two Dimensions

Gradient descent in one dimension is straightforward enough. Let's move to two dimensions to increase our intuition about the algorithm. The code referenced below is in the file *gd_2d.py*. We'll first consider the case where the function has a single minimum, then look at cases with multiple minima.

Gradient Descent with a Single Minimum

To work in two dimensions, we need a scalar function of a vector, $f(\boldsymbol{x}) = f(x, y)$, where, to make it easier to follow, we separate the vector into its components, $\boldsymbol{x} = (x, y)$.

The first function we'll work with is

$$f(x, y) = 6x^2 + 9y^2 - 12x - 14y + 3$$

To implement gradient descent, we need the partial derivatives as well:

$$\frac{\partial f}{\partial x} = 12x - 12$$

$$\frac{\partial f}{\partial y} = 18y - 14$$

Our update equations become

$$x \leftarrow x - \eta\, \frac{\partial f}{\partial x} = x - \eta\, (12x - 12)$$

$$y \leftarrow y - \eta\, \frac{\partial f}{\partial y} = y - \eta\, (18y - 14)$$

In code, we define the function and partial derivatives:

```
def f(x,y):
    return 6*x**2 + 9*y**2 - 12*x - 14*y + 3
def dx(x):
    return 12*x - 12
def dy(y):
    return 18*y - 14
```

Since the partial derivatives are independent of the other variable, we get away with passing only *x* or *y*. We'll see an example later in this section where that's not the case.

Gradient descent follows the same pattern as before: select an initial position, this time a vector, iterate for some number of steps, and plot the path. The function is 2D, so we first plot it using contours, as shown next.

```
N = 100
x,y = np.meshgrid(np.linspace(-1,3,N), np.linspace(-1,3,N))
z = f(x,y)
plt.contourf(x,y,z,10, cmap="Greys")
plt.contour(x,y,z,10, colors='k', linewidths=1)
plt.plot([0,0],[-1,3],color='k',linewidth=1)
plt.plot([-1,3],[0,0],color='k',linewidth=1)
plt.plot(1,0.7777778,color='k',marker='+')
```

This code requires some explanation. To plot contours, we need a representation of the function over a grid of (x, y) pairs. To generate the grid, we use NumPy, specifically np.meshgrid. The arguments to np.meshgrid are the x and y points, here provided by np.linspace, which itself generates a vector from -1 to 3 of $N = 100$ evenly spaced values. The np.meshgrid function returns two 100×100 matrices. The first contains the x values over the given range, and the second contains the y values. All possible (x, y) pairs are represented in the return value to form a grid of points covering the region of $-1 \ldots 3$ in both x and y. Passing these points to the function then returns z, a 100×100 matrix of the function value at each (x, y) pair.

We could plot the function in 3D, but that's difficult to see and unnecessary in this case. Instead, we'll use the function values in x, y, and z to generate contour plots. Contour plots show 3D information as a series of lines of equal z value. Think of lines around a hill on a topographic map, where each line is at the same altitude. As the hill gets higher, the lines enclose successively smaller regions.

Contour plots come in two varieties, as either lines of equal function value or shading over ranges of the function. We'll plot both varieties using a grayscale map. That's the net result of calling Matplotlib's plt.contourf and plt.contour functions. The remaining plt.plot calls show the axes and mark the function minimum with a plus sign. The contour plots are such that lighter shades imply lower function values.

We're now ready to plot the sequence of gradient descent steps. We'll plot each position in the sequence and connect them with a dashed line to make the path clear (see Listing 11-1). In code, that's

```
x = xold = -0.5
y = yold = 2.9
for i in range(12):
    plt.plot([xold,x],[yold,y], marker='o', linestyle='dotted', color='k')
    xold = x
    yold = y
    x = x - 0.02 * dx(x)
    y = y - 0.02 * dy(y)
```

Listing 11-1: Gradient descent in two dimensions

We begin at $(x, y) = (-0.5, 2.9)$ and take 12 gradient descent steps. To connect the last position to the new position using a dashed line, we track both the current position in x and y and the previous position, (x_{old}, y_{old}). The gradient descent step updates both x and y using $\eta = 0.02$ and calling the respective partial derivative functions, dx and dy.

Figure 11-3 shows the gradient descent path that Listing 11-1 follows (circles) along with two other paths starting at $(1.5, -0.8)$ (squares) and $(2.7, 2.3)$ (triangles).

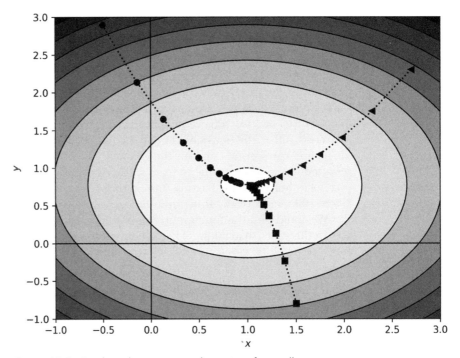

Figure 11-3: Gradient descent in two dimensions for small steps

All three gradient descent paths converge toward the minimum of the function. This isn't surprising, as the function has only one minimum. If the function has a single minimum, then gradient descent will eventually find it. If the step size is too small, many steps might be necessary, but they will ultimately converge on the minimum. If the step size is too large, gradient descent may oscillate around the minimum but continually step over it.

Let's change our function a bit to stretch it in the x direction relative to the y direction:

$$f(x, y) = 6x^2 + 40y^2 - 12x - 30y + 3$$

This function has partials $\partial f/\partial x = 12x - 12$ and $\partial f/\partial y = 80y - 30$.

Additionally, let's pick two starting locations, $(-0.5, 2.3)$ and $(2.3, 2.3)$, and generate a sequence of gradient descent steps with $\eta = 0.02$ and $\eta = 0.01$, respectively. Figure 11-4 shows the resulting paths.

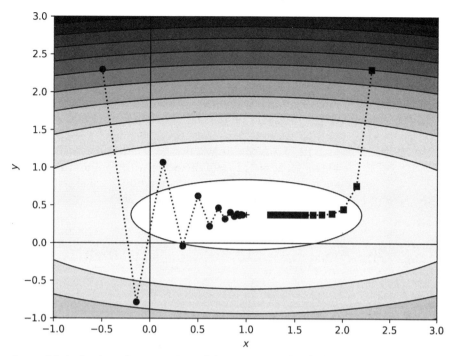

Figure 11-4: Gradient descent in 2D with larger steps and a slightly different function

Consider the $\eta = 0.02$ (circle) path first. The new function is like a canyon, narrow in y but long in x. The larger step size oscillates up and down in y as it moves toward the minimum in x. Bouncing off the canyon walls aside, we still find the minimum.

Now, take a look at the $\eta = 0.01$ (square) path. It quickly falls into the canyon and then moves slowly over the flat region along the canyon floor toward the minimum position. The component of the vector gradient (the x and y partial derivative values) along the x direction is small in the canyon, so motion along x is proportionately slow. There is no motion in the y direction—the canyon is steep, and the relatively small learning rate has already located the canyon floor, where the gradient is primarily along x.

What's the lesson here? Again, the step size matters. However, the shape of the function matters even more. The minimum of the function lies at the bottom of a long, narrow canyon. The gradient along the canyon is tiny; the canyon floor is flat in the x direction, so motion is slow because it depends on the gradient value. We frequently encounter this effect in deep learning: if the gradient is small, learning is slow. This is why the rectified linear unit has come to dominate deep learning; the gradient is a constant one for positive inputs. For a sigmoid or hyperbolic tangent, the gradient approaches zero when inputs are far from zero.

Gradient Descent with Multiple Minima

The functions we've examined so far have a single minimum value. What if that isn't the case? Let's see what happens to gradient descent when the function has more than one minimum. Consider this function:

$$f(x, y) = -2 \exp\left(-\frac{1}{2}\left((x+1)^2 + (y-1)^2\right)\right) - \exp\left(-\frac{1}{2}\left((x-1)^2 + (y+1)^2\right)\right) \qquad (11.4)$$

Equation 11.4 is the sum of two inverted Gaussians, one with a minimum value of -2 at $(-1, 1)$ and the other with a minimum of -1 at $(1, -1)$. If gradient descent is to find the global minimum, it should find it at $(-1, 1)$. The code for this example is in *gd_multiple.py*.

The partial derivatives are

$$\frac{\partial f}{\partial x} = 2(x+1) \exp\left(-\frac{1}{2}\left((x+1)^2 + (y-1)^2\right)\right) + (x-1) \exp\left(-\frac{1}{2}\left((x-1)^2 + (y+1)^2\right)\right)$$

$$\frac{\partial f}{\partial y} = 2(y-1) \exp\left(-\frac{1}{2}\left((x+1)^2 + (y-1)^2\right)\right) + (y+1) \exp\left(-\frac{1}{2}\left((x-1)^2 + (y+1)^2\right)\right)$$

which translates into the following code:

```
def f(x,y):
    return -2*np.exp(-0.5*((x+1)**2+(y-1)**2)) + \
           -np.exp(-0.5*((x-1)**2+(y+1)**2))

def dx(x,y):
    return 2*(x+1)*np.exp(-0.5*((x+1)**2+(y-1)**2)) + \
           (x-1)*np.exp(-0.5*((x-1)**2+(y+1)**2))

def dy(x,y):
    return (y+1)*np.exp(-0.5*((x-1)**2+(y+1)**2)) + \
           2*(y-1)*np.exp(-0.5*((x+1)**2+(y-1)**2))
```

Notice, in this case, the partial derivatives do depend on both x and y.

The code for the gradient descent portion of *gd_multiple.py* is as before. Let's run the cases in Table 11-1.

Table 11-1: Different Starting Positions and Number of Gradient Descent Steps Taken

Starting point	Steps	Symbol
(-1.5,1.2)	9	circle
(1.5,-1.8)	9	square
(0,0)	20	plus
(0.7,-0.2)	20	triangle
(1.5,1.5)	30	asterisk

The Symbol column refers to the plot symbol used in Figure 11-5. For all cases, $\eta = 0.4$.

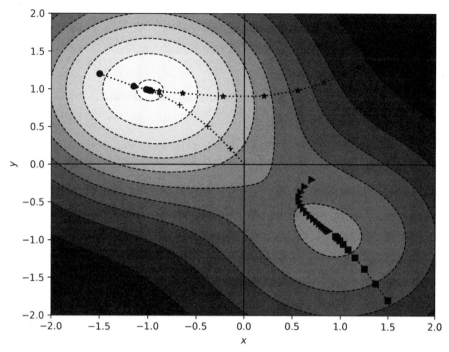

Figure 11-5: Gradient descent for a function with two minima

The gradient descent paths indicated in Figure 11-5 make sense. In three of the five cases, the path does move into the well that the deeper of the two minima defines—a successful search. However, for the triangle and the square, gradient descent fell into the wrong minimum. Clearly, how successful gradient descent is, in this case, depends on where we start the process. Once the path moves downhill to a deeper position, gradient descent has no way to escape upward to find a potentially better minimum.

Current thinking is that the loss landscape for a deep learning model contains many minima. It's also currently believed that in most cases, the minima are pretty similar, which partially explains the success of deep learning models—to train them, you don't need to find the one, magic, global minimum of the loss, only one of the (probably) many that are (probably) about as good as any of the others.

I selected the initial positions used for the examples in this section intentionally based on knowledge of the function's form. For a deep learning model, picking the starting point means random initialization of the weights and biases. In general, we don't know the form of the loss function, so initialization is a shot in the dark. Most of the time, or at least much of the time, gradient descent produces a well-performing model. Sometimes, however, it doesn't; it fails miserably. In those cases, it's possible the initial position was like the square in Figure 11-5: it fell into an inferior local minimum because it started in a bad place.

Now that we have a handle on gradient descent, what it is, and how it works, let's investigate how we can apply it in deep learning.

Stochastic Gradient Descent

Training a neural network is primarily the act of minimizing the loss function while preserving generalizability via various forms of regularization. In Chapter 10, we wrote the loss as $L(\theta; x, y)$ for a vector of the weights and biases, θ (theta), and training instances (x, y), where x is the input vectors and y is the known labels. Note how here, x is a stand-in for *all* training data, not just a single sample.

Gradient descent needs $\partial L / \partial \theta$, which we get via backpropagation. The expression $\partial L / \partial \theta$ is a concise way of referring to all the individual weight and bias error terms backpropagation gives us. We get $\partial L / \partial \theta$ by averaging the error over the training data. This begs the question: Do we average over all of the training data or only some of the training data?

Passing all the training data through the model before taking a gradient descent step is called batch training. At first blush, batch training seems sensible. After all, if our training set is a good sample from the parent distribution that generates the sort of data our model intends to work with, then why not use all of that sample to do gradient descent?

When datasets were small, batch training was the natural thing to do. However, models got bigger, as did datasets, and suddenly the computational burden of passing *all* the training data through the model for each gradient descent step became too much. This chapter's examples already hint that many gradient descent steps might be necessary to find a good minimum position, especially for tiny learning rates.

Therefore, practitioners began to use subsets of the training data for each gradient descent step—the *minibatch*. Minibatch training was probably initially viewed as a compromise, as the gradient calculated over the minibatch was "wrong" because it wasn't based on the performance of the full training set.

Of course, the difference between *batch* and *minibatch* is just an agreed-upon fiction. In truth, it's a continuum from a minibatch of one to a minibatch of all available samples. With that in mind, all the gradients computed during network training are "wrong," or at least incomplete, as they are based on incomplete knowledge of the data generator and the full set of data it could generate.

Rather than a concession, then, minibatch training is reasonable. The gradient over a small minibatch is noisy compared to that computed over a larger minibatch, in the sense that the small minibatch gradient is a coarser estimate of the "real" gradient. When things are noisy or random, the word *stochastic* tends to show up, as it does here. Gradient descent with minibatches is *stochastic gradient descent (SGD)*.

In practice, gradient descent using smaller minibatches often leads to models that perform better than those trained with larger minibatches. The rationale generally given is that the noisy gradient of the smaller minibatch helps gradient descent avoid falling into poor local minima of the loss landscape. We saw this effect in Figure 11-5, where the triangle and the square both fell into the wrong minimum.

Again, we find ourselves strangely fortunate. Before, we were fortunate because first-order gradient descent succeeded in training models that shouldn't train due to nonlinear loss landscapes, and now we get a boost by intentionally using small amounts of data to estimate gradients, thereby skipping a computational burden likely to make the entire enterprise of deep learning too cumbersome to implement in many cases.

How large should our minibatch be? Minibatch size is a *hyperparameter*, something we need to select to train the model, but is not part of the model itself. The proper minibatch size is dataset-dependent. For example, in the extreme, we could take a gradient descent step for each sample, which sometimes works well. This case is often referred to as *online learning*. However, especially if we use layers like batch normalization, we need a minibatch large enough to make the calculated means and standard deviations reasonable estimates. Again, as with most everything else in deep learning at present, it's empirical, and you need to both have intuition and try many variations to optimize the training of the model. This is why people work on *AutoML* systems, systems that seek to do all the hyperparameter tuning for you.

Another good question: What should be in the minibatch? That is, what small subset of the full dataset should we use? Typically, the order of the samples in the training set is randomized, and minibatches are pulled from the set as successive chunks of samples until all samples have been used. Using all the samples in the dataset defines one epoch, so the number of samples in the training set divided by the minibatch size determines the number of minibatches per epoch.

Alternatively, as we did for *NN.py*, a minibatch might genuinely be a random sampling from the available data. It's possible that a particular training sample is never used while another is used many times, but on the whole, the majority of the dataset is used during training.

Some toolkits train for a specified number of minibatches. Both *NN.py* and Caffe operate this way. Other toolkits, like Keras and sklearn, use epochs. Gradient descent steps happen after a minibatch is processed. Larger minibatches result in fewer gradient descent steps per epoch. To compensate, practitioners using toolkits that use epochs need to ensure that the number of gradient descent steps increases as minibatch size increases—larger minibatches require more epochs to train well.

To recap, deep learning does not use full batch training for at least the following reasons:

1. The computational burden is too great to pass the entire training set through the model for each gradient descent step.
2. The gradient computed from the average loss over a minibatch is a noisy but reasonable estimate of the true, and ultimately unknowable, gradient.
3. The noisy gradient points in a slightly wrong direction in the loss landscape, thereby possibly avoiding bad minima.
4. Minibatch training simply works better in practice for many datasets.

Reason #4 should not be underestimated: many practices in deep learning are employed initially because they simply work better. Only later are they justified by theory, if at all.

As we already implemented SGD in Chapter 10 (see *NN.py*), we won't reimplement it here, but in the next section, we'll add momentum to see how that affects neural network training.

Momentum

Vanilla gradient descent relies solely on the value of the partial derivative multiplied by the learning rate. If the loss landscape has many local minima, especially if they're steep, vanilla gradient descent might fall into one of the minima and be unable to recover. To compensate, we can modify vanilla gradient descent to include a *momentum* term, a term that uses a fraction of the previous step's update. Including this momentum in gradient descent adds inertia to the algorithm's motion through the loss landscape, thereby potentially allowing gradient descent to move past bad local minima.

Let's define and then experiment with momentum using 1D and 2D examples, as we did earlier. After that, we'll update our *NN.py* toolkit to use momentum to see how that affects models trained on more complex datasets.

What Is Momentum?

In physics, the momentum of a moving object is defined as the mass times the velocity, $p = mv$. However, velocity itself is the first derivative of the position, $v = dx/dt$, so momentum is mass times how fast the position of the object is changing in time.

For gradient descent, *position* is the function value, and *time* is the argument to the function. The *velocity*, then, is how fast the function value changes with a change in the argument, $\partial f / \partial x$. Therefore, we can think of *momentum* as a scaled velocity term. In physics, the scale factor is the mass. For gradient descent, the scale factor is μ (mu), a number between zero and one.

If we call the gradient including the momentum term v, then the gradient descent update equation that was

$$x \leftarrow x - \eta \, \frac{\partial f}{\partial x}$$

becomes

$$v \leftarrow \mu v - \eta \, \frac{\partial f}{\partial x}$$

$$x \leftarrow x + v \tag{11.5}$$

for some initial velocity, $v = 0$, and the "mass," μ.

Let's walk through Equation 11.5 to understand what it means. The two-step update, first v and then x, makes it easy to iterate, as we know we must do for gradient descent. If we substitute v into the update equation for x, we get

$$x \leftarrow x + \mu v - \eta \, \frac{\partial f}{\partial x}$$

This makes it clear that the update includes the gradient step we had previously but adds back in a fraction of the previous step size. It's a fraction because we restrict μ to $[0, 1]$. If $\mu = 0$, we're back to vanilla gradient descent. It might be helpful to think of μ as a scale factor, the fraction of the previous velocity to keep along with the current gradient value.

The momentum term tends to keep motion through the loss landscape heading in its previous direction. The value of μ determines the strength of that tendency. Deep learning practitioners typically use $\mu = 0.9$, so most of the previous update direction is maintained in the next step, with the current gradient providing a small adjustment. Again, like many things in deep learning, this number was chosen empirically.

Newton's first law of motion states that an object in motion remains in motion unless acted upon by an outside force. Resistance to an external force is related to the object's mass and is called *inertia*. So, we might also view the μv term as inertia, which might have been a better name for it.

Regardless of the name, now that we have it, let's see what it does to the 1D and 2D examples we worked through earlier using vanilla gradient descent.

Momentum in 1D

Let's modify the 1D and 2D examples above to use a momentum term. We'll start with the 1D case. The updated code is in the file *gd_1d_momentum.py* and appears here as Listing 11-2.

```
import matplotlib.pylab as plt

def f(x):
    return 6*x**2 - 12*x + 3
def d(x):
    return 12*x - 12

❶ m = ['o','s','>','<','*','+','p','h','P','D']
  x = np.linspace(0.75,1.25,1000)
  plt.plot(x,f(x))

❷ x = xold = 0.75
  eta = 0.09
  mu = 0.8
  v = 0.0

  for i in range(10):
❸     plt.plot([xold,x], [f(xold),f(x)], marker=m[i], linestyle='dotted',
          color='r')
      xold = x
      v = mu*v - eta * d(x)
      x = x + v

  for i in range(40):
      v = mu*v - eta * d(x)
      x = x + v

❹ plt.plot(x,f(x),marker='X', color='k')
```

Listing 11-2: Gradient descent in one dimension with momentum

Listing 11-2 is a bit dense, so let's parse it out. First, we are plotting, so we include Matplotlib. Next, we define the function, f(x), and its derivative, d(x), as we did before. To configure plotting, we define a collection of markers ❶ and then plot the function itself. As before, we begin at $x = 0.75$ ❷ and set the step size (eta), momentum (mu), and initial velocity (v).

We're now ready to iterate. We'll use two gradient descent loops. The first plots each step ❸ and the second continues gradient descent to demonstrate that we do eventually locate the minimum, which we mark with an 'X' ❹. For each step, we calculate the new velocity by mimicking Equation 11.5, and then we add the velocity to the current position to get the next position.

Figure 11-6 shows the output of *gd_1d_momentum.py*.

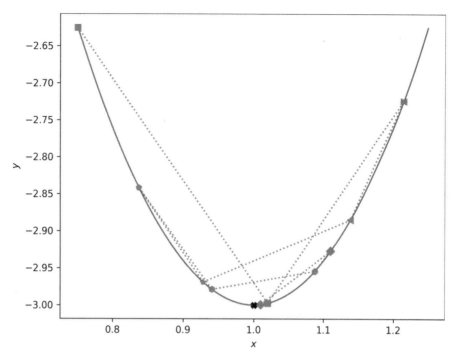

Figure 11-6: Gradient descent in one dimension with momentum

Note that we intentionally used a large step size (η), so we overshoot the minimum. The momentum term tends to overshoot minima as well. If you follow the dashed line and the sequence of plot markers, you can walk through the first 10 gradient descent steps. There is oscillation, but the oscillation is damped and eventually settles at the minimum, as marked. Adding momentum enhanced the overshoot due to the large step size. However, even with the momentum term, which isn't advantageous here, because there's only one minimum, with enough gradient descent steps, we find the minimum in the end.

Momentum in 2D

Now, let's update our 2D example. We're working with the code in *gd _momentum.py*. Recall that, for the 2D example, the function is the sum of two inverted Gaussians. Including momentum updates the code slightly, as shown in Listing 11-3:

```
def gd(x,y, eta,mu, steps, marker):
    xold = x
    yold = y
❶   vx = vy = 0.0
    for i in range(steps):
        plt.plot([xold,x],[yold,y], marker=marker,
                 linestyle='dotted', color='k')
```

```
    xold = x
    yold = y
❷  vx = mu*vx - eta * dx(x,y)
    vy = mu*vy - eta * dy(x,y)
❸  x = x + vx
    y = y + vy

❹ gd( 0.7,-0.2, 0.1,  0.9, 25, '>')
  gd( 1.5, 1.5, 0.02, 0.9, 90, '*')
```

Listing 11-3: Gradient descent in two dimensions with momentum

Here, we have the new function, gd, which performs gradient descent with momentum beginning at (x,y), using the given μ and η, and runs for steps iterations.

The initial velocity is set ❶, and the loop begins. The velocity update of Equation 11.5 becomes vx = mu*vx - eta * dx(x,y) ❷, and the position update becomes x = x + vx ❸. As before, a line is plotted between the last position and the current one to track motion through the function landscape.

The code in *gd_momentum.py* traces the motion starting at two of the points we used before, $(0.7, -0.2)$ and $(1.5, 1.5)$ ❹. Note the number of steps and learning rate vary by point to keep the plot from becoming too cluttered. The output of *gd_momentum.py* is Figure 11-7.

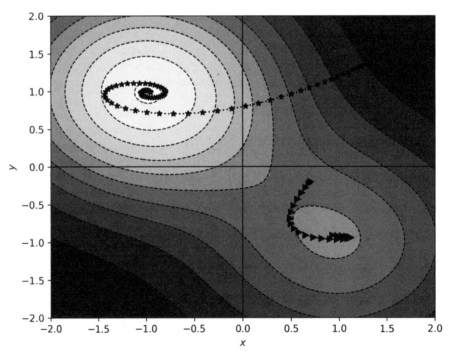

Figure 11-7: Gradient descent in two dimensions with momentum

Compare the paths in Figure 11-7 with those in Figure 11-5. Adding momentum has pushed the paths, so they tend to keep moving in the same direction. Notice how the path beginning at $(1.5, 1.5)$ spirals toward the minimum, while the other path curves toward the shallower minimum, passes it, and backtracks toward it again.

The momentum term alters the dynamics of motion through the function space. However, it's not immediately evident that momentum adds anything helpful. After all, the $(1.5, 1.5)$ starting position using vanilla gradient descent moved directly to the minimum position without spiraling.

Let's add momentum to our *NN.py* toolkit and see if it buys us anything when training real neural networks.

Training Models with Momentum

To support momentum in *NN.py*, we need to tweak the `FullyConnectedLayer` method in two places. First, as shown in Listing 11-4, we modify the constructor to allow a `momentum` keyword:

```
def __init__(self, input_size, output_size, momentum=0.0):
    self.delta_w = np.zeros((input_size, output_size))
    self.delta_b = np.zeros((1,output_size))
    self.passes = 0
    self.weights = np.random.rand(input_size, output_size) - 0.5
    self.bias = np.random.rand(1, output_size) - 0.5
❶   self.vw = np.zeros((input_size, output_size))
    self.vb = np.zeros((1, output_size))
    self.momentum = momentum
```

Listing 11-4: Adding the momentum keyword

Here, we add a `momentum` keyword, with a default of zero, into the argument list. Then, we define initial velocities for the weights (`vw`) and biases (`vb`) ❶. These are matrices of the proper shape initialized to zero. We also keep the momentum argument for later use.

The second modification is to the step method, as Listing 11-5 shows:

```
def step(self, eta):
❶   self.vw = self.momentum * self.vw - eta * self.delta_w / self.passes
    self.vb = self.momentum * self.vb - eta * self.delta_b / self.passes
❷   self.weights = self.weights + self.vw
    self.bias = self.bias + self.vb
    self.delta_w = np.zeros(self.weights.shape)
    self.delta_b = np.zeros(self.bias.shape)
    self.passes = 0
```

Listing 11-5: Updating the step to include momentum

We implement Equation 11.5, first for the weights ❶, then for the biases in the line after. We multiply the momentum (μ) by the previous velocity, then subtract the average error over the minibatch, multiplied by the learning rate. We then move the weights and biases by adding the velocity ❷. That's all we need to do to incorporate momentum. Then, to use it, we add the momentum keyword to each fully connected layer when building the network, as shown in Listing 11-6:

```
net = Network()
net.add(FullyConnectedLayer(14*14, 100, momentum=0.9))
net.add(ActivationLayer())
net.add(FullyConnectedLayer(100, 50, momentum=0.9))
net.add(ActivationLayer())
net.add(FullyConnectedLayer(50, 10, momentum=0.9))
net.add(ActivationLayer())
```

Listing 11-6: Specifying momentum when building the network

Adding momentum per layer opens up the possibility of using layer-specific momentum values. While I'm unaware of any research doing so, it seems a fairly obvious thing to try, so by now, someone has likely experimented with it. For our purposes, we'll set the momentum of all layers to 0.9 and move on.

How should we test our new momentum? We could use the MNIST dataset we used above, but it's not a good candidate, because it's too easy. Even a simple fully connected network achieves better than 97 percent accuracy. Therefore, we'll replace the MNIST digits dataset with another, similar dataset that's known to be more of a challenge: the Fashion-MNIST dataset. (See "Fashion-MNIST: A Novel Image Dataset for Benchmarking Machine Learning Algorithms" by Han Xiao et al., arXiv:1708.07747 [2017].)

The *Fashion-MNIST dataset (FMNIST)* is a drop-in replacement for the existing MNIST dataset. It contains images from 10 classes of clothing, all 28×28-pixel grayscale. For our purposes, we'll do as we did for MNIST and reduce the 28×28-pixel images to 14×14 pixels. The images are in the dataset directory as NumPy arrays. Let's train a model using them. The code for the model is similar to that of Listing 10-7, except in Listing 11-7 we replace the MNIST dataset with FMNIST:

```
x_train = np.load("fmnist_train_images_small.npy")/255
x_test  = np.load("fmnist_test_images_small.npy")/255
y_train = np.load("fmnist_train_labels_vector.npy")
y_test  = np.load("fmnist_test_labels.npy")
```

Listing 11-7: Loading the Fashion-MNIST dataset

We also include code to calculate the Matthews correlation coefficient (MCC) on the test data. We first encountered the MCC in Chapter 4, where we learned that it's a better measure of a model's performance than the accuracy is. The code to run is in *fmnist.py*. Taking around 18 minutes on an older Intel i5 box, one run of it produced

```
[[866   1  14  28   8   1  68   0  14   0]
 [  5 958   2  25   5   0   3   0   2   0]
 [ 20   1 790  14 126   0  44   1   3   1]
 [ 29  21  15 863  46   1  20   0   5   0]
```

```
[  0   0  91  22 849   1  32   0   5   0]
[  0   0   0   1   0 960   0  22   2  15]
[161   2 111  38 115   0 556   0  17   0]
[  0   0   0   0   0  29   0 942   0  29]
[  1   0   7   5   6   2   2   4 973   0]
[  0   0   0   0   0   6   0  29   1 964]]
```

accuracy = 0.8721000
MCC = 0.8584048

The confusion matrix, still 10×10 because of the 10 classes in FM-NIST, is quite noisy compared to the very clean confusion matrix we saw with MNIST proper. This is a challenging dataset for fully connected models. Recall that the MCC is a measure where the closer it is to one, the better the model.

The confusion matrix above is for a model trained without momentum. The learning rate was 1.0, and it was trained for 40,000 minibatches of 64 samples. What happens if we add momentum of 0.9 to each fully connected layer and reduce the learning rate to 0.2? When we add momentum, it makes sense to reduce the learning rate so we aren't taking large steps compounded by the momentum already moving in a particular direction. Do explore what happens if you run *fmnist.py* with a learning rate of 0.2 and no momentum.

The version of the code with momentum is in *fmnist_momentum.py*. After about 20 minutes, one run of this code produced

```
[[766   5  14  61   2   1 143   0   8   0]
 [  1 958   2  30   3   0   6   0   0   0]
 [ 12   0 794  16  98   0  80   0   0   0]
 [  8  11  13 917  21   0  27   0   3   0]
 [  0   0  84  44 798   0  71   0   3   0]
 [  0   0   0   1   0 938   0  31   1  29]
 [ 76   2  87  56  60   0 714   0   5   0]
 [  0   0   0   0   0  11   0 963   0  26]
 [  1   1   6   8   5   1  10   4 964   0]
 [  0   0   0   0   0   6   0  33   0 961]]
```

accuracy = 0.8773000
MCC = 0.8638721

---

giving us a slightly higher MCC. Does that mean momentum helped? Maybe. As we well understand by now, training neural networks is a stochastic process. So, we can't rely on results from a single training of the models. We need to train the models many times and perform statistical tests on the results. Excellent! This gives us a chance to put the hypothesis testing knowledge we gained in Chapter 4 to good use.

Instead of running *fmnist.py* and *fmnist_momentum.py* one time each, let's run them 22 times each. This takes the better part of a day on my old Intel i5 system, but patience is a virtue. The net result is 22 MCC values for the model with momentum and 22 for the model without momentum. There's nothing magical about 22 samples, but we intend to use the Mann-Whitney U test, and the rule of thumb for that test is to have at least 20 samples in each dataset.

Figure 11-8 displays histograms of the results.

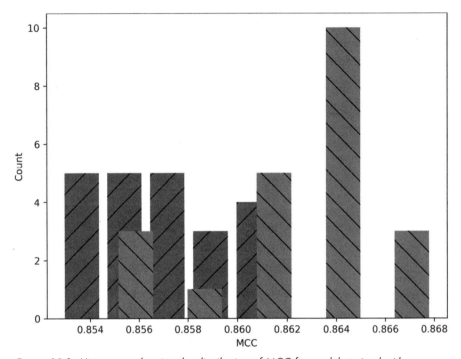

Figure 11-8: Histograms showing the distribution of MCC for models trained with momentum (light gray) and without (dark gray)

The darker gray bars are the no-momentum MCC values, and the lighter bars are those with momentum. Visually, the two are largely distinct from each other. The code producing Figure 11-8 is in the file *fmnist_analyze.py*. Do take a look at the code. It uses SciPy's ttest_ind and mannwhitneyu along with the implementation we gave in Chapter 4 of Cohen's *d* to calculate the effect size. The MCC values themselves are in the NumPy files listed in the code.

Along with the graph, *fmnist_analyze.py* produces the following output:

```
no momentum: 0.85778 +/- 0.00056
momentum   : 0.86413 +/- 0.00075

t-test momentum vs no (t,p): (6.77398299, 0.00000003)
Mann-Whitney U            : (41.00000000, 0.00000126)
Cohen's d                 : 2.04243
```

where the top two lines are the mean and the standard error of the mean. The t-test results are (t, p), the *t*-test statistic and associated *p*-value. Similarly, the Mann-Whitney U test results are (U, p), the U statistic and its p-value. Recall how the Mann-Whitney U test is a nonparametric test assuming nothing about the shape of the distribution of MCC values. The t-test assumes they are normally distributed. As we have only 22 samples each, we really can't make any definitive statement about whether the results are normally distributed; the histograms don't look much like Gaussian curves. That's why we included the Mann-Whitney U test results.

A glance at the respective p-values tells us that the difference in means between the MCC values with and without momentum is highly statistically significant in favor of the with-momentum results. The t-value is positive, and the with-momentum result was the first argument. What of Cohen's d-value? It's a bit above 2.0, indicating a (very) large effect size.

Can we *now* say that momentum helps in this case? Probably. It produced better performing models given the hyperparameters we used. The stochastic nature of training neural networks makes it possible that we could tweak the hyperparameters of both models to eliminate the difference we see in the data we have. The architecture between the two is fixed, but nothing says the learning rate and minibatch size are optimized for either model.

A punctilious researcher would feel compelled to run an optimization process over the hyperparameters and, once satisfied that they'd found the very best model for both approaches, make a more definite statement after repeating the experiment. We, thankfully, are not punctilious researchers. Instead, we'll use the evidence we have, along with the several decades of wisdom acquired by the world's machine learning researchers regarding the utility of momentum in gradient descent, to state that, yes, momentum helps models learn, and you should use it in most cases.

However, the normality question is begging for further investigation. We are, after all, seeking to improve our mathematical *and* practical intuition regarding deep learning. Therefore, let's train the with-momentum model for FMNIST, not 22 times but 100 times. As a concession, we'll reduce the number of minibatches from 40,000 to 10,000. Still, expect to spend the better part of a day waiting for the program to finish. The code, which we won't walk through here, is in *fmnist_repeat.py*.

Figure 11-9 presents a histogram of the results.

Clearly, this distribution does not look at all like a normal curve. The output of *fmnist_repeat.py* includes the result of SciPy's normaltest function. This function performs a statistical test on a set of data under the null hypothesis that the data *is* normally distributed. Therefore, a p-value below, say, 0.05 or 0.01, indicates data that is not normally distributed. Our p-value is virtually zero.

What to make of Figure 11-9? First, as the results are certainly not normal, we aren't justified in using a t-test. However, we also used the nonparametric Mann-Whitney U test and found highly statistically significant results, so our claims above are still valid. Second, the long tail of the distribution in Figure 11-9 is to the left. We might even make an argument that the result is possibly bimodal: that there are two peaks, one near 0.83 and the other, smaller one near an MCC of 0.75.

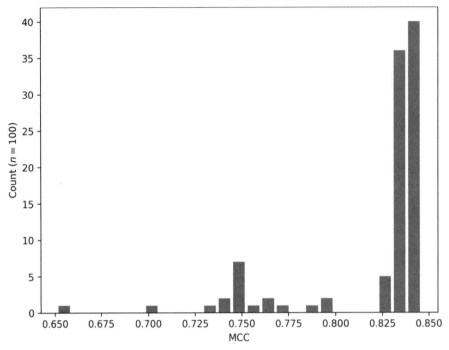

Figure 11-9: Distribution of MCC values for 100 trainings of the FMNIST model

Most models trained to a relatively consistent level of performance, with an MCC near 0.83. However, the long tail indicates that when the model wasn't reasonably good, it was just plain horrid.

Intuitively, Figure 11-9 seems reasonable to me. We know stochastic gradient descent is susceptible to improper initialization, and our little toolkit is using old-school small random value initialization. It seems likely that we have an increased chance of starting at a poor location in the loss landscape and are doomed after that to poor performance.

What if the tail were on the right? What might that indicate? A long tail on the right would mean most model performance is mediocre to poor, but, on occasion, an especially "bright" model comes along. Such a scenario would mean that better models are out there, but that our training and/or initialization strategy isn't particularly good at finding them. I think the tail on the left is preferable—most models find reasonably good local minima, so most trainings, unless horrid, end up in pretty much the same place in terms of performance.

Now, let's examine a common variant of momentum, one that you'll no doubt run across during your sojourn through deep learning.

Nesterov Momentum

Many deep learning toolkits include the option to use *Nesterov momentum* during gradient descent. Nesterov momentum is a modification of gradient descent widely used in the optimization community. The version typically implemented in deep learning updates standard momentum from

$$v \leftarrow \mu v - \eta \, \nabla f(x)$$

$$x \leftarrow x + v$$

to

$$v \leftarrow \mu v - \eta \, \nabla f(x + \mu v)$$

$$x \leftarrow x + v \qquad\qquad (11.6)$$

where we're using gradient notation instead of partials of a loss function to indicate that the technique is general and applies to any function, $f(x)$.

The difference between standard momentum and deep learning Nesterov momentum is subtle, just a term that's added to the argument of the gradient. The idea is to use the existing momentum to calculate the gradient, not at the current position, x, but the position gradient descent would be at if it continued further using the current momentum, $x + \mu v$. We then use the gradient's value at that position to update the current position, as before.

The claim, well demonstrated for optimization in general, is that this tweak leads to faster convergence, meaning gradient descent will find the minimum in fewer steps. However, even though toolkits implement it, there is reason to believe the noise that stochastic gradient descent with minibatches introduces offsets the adjustment to the point where it's unlikely Nesterov momentum is any more useful for training deep learning models than regular momentum. (For more on this, see the comment on page 292 of *Deep Learning* by Ian Goodfellow et al.)

However, the 2D example in this chapter uses the actual function to calculate gradients, so we might expect Nesterov momentum to be effective in that case. Let's update the 2D example, minimizing the sum of two inverted Gaussians, and see if Nesterov momentum improves convergence, as claimed. The code we'll run is in *gd_nesterov.py* and is virtually identical to the code in *gd_momentum.py*. Additionally, I tweaked both files a tiny bit to return the final position after gradient descent is complete. That way, we can see how close we are to the known minima.

Implementing Equation 11.6 is straightforward and affects only the velocity update, causing

```
vx = mu*vx - eta * dx(x,y)
vy = mu*vy - eta * dy(x,y)
```

to become

```
vx = mu * vx - eta * dx(x + mu * vx,y)
vy = mu * vy - eta * dy(x,y + mu * vy)
```

to add the momentum for each component, x and y. Everything else remains the same.

Figure 11-10 compares standard momentum (top, from Figure 11-7) and Nesterov momentum (bottom).

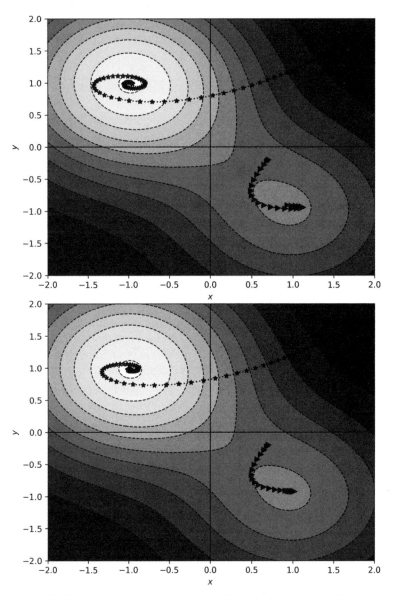

Figure 11-10: Standard momentum (top) and Nesterov momentum (bottom)

Visually, Nesterov momentum shows less of an overshoot, especially for the spiral marking the path beginning at $(1.5, 1.5)$. What about the final location that each approach returns? We get Table 11-2.

Table 11-2: Final Location for Gradient Descent With and Without Nesterov Momentum

Initial point	Standard	Nesterov	Minimum
(1.5,1.5)	(−0.9496, 0.9809)	(−0.9718, 0.9813)	(−1,1)
(0.7,−0.2)	(0.8807, −0.9063)	(0.9128, −0.9181)	(1,−1)

The Nesterov momentum results are closer to the known minima than the standard momentum results after the same number of gradient descent steps.

Adaptive Gradient Descent

The gradient descent algorithm is almost trivial, which invites adaptation. In this section, we'll walk through the math behind three variants of gradient descent popular with the deep learning community: RMSprop, Adagrad, and Adam. Of the three, Adam is the most popular by far, but the others are well worth understanding, as they build in succession leading up to Adam. All three of these algorithms adapt the learning rate on the fly in some manner.

RMSprop

Geoffrey Hinton introduced *RMSprop*, which stands for *root mean square propagation*, in his 2012 Coursera lecture series. Much like momentum (with which it can be combined), RMSprop is gradient descent that tracks the value of the gradient as it changes and uses that value to modify the step taken.

RMSprop uses a *decay term*, γ (gamma), to calculate a running average of the gradients as the algorithm progresses. In his lecture, Hinton uses $\gamma = 0.9$.

The gradient descent update becomes

$$m \leftarrow \gamma m + (1 - \gamma)[\nabla f(\boldsymbol{x})]^2$$

$$\boldsymbol{v} \leftarrow -\frac{\eta}{\sqrt{m}} \nabla f(\boldsymbol{x})$$

$$\boldsymbol{x} \leftarrow \boldsymbol{x} + \boldsymbol{v} \tag{11.7}$$

First, we update m, the running average of the squares of the gradients, weighted by γ, the decay term. Next comes the velocity term, which is almost the same as in vanilla gradient descent, but we divide the learning rate by the running average's square root, hence the RMS part of RMSprop. We then subtract the scaled velocity from the current position to take the step. We're writing the step as an addition, similar to the momentum equations above (Equations 11.5 and 11.6); note the minus sign before the velocity update.

RMSprop works with momentum as well. For example, extending RMSprop with Nesterov momentum is straightforward:

$$m \leftarrow \gamma m + (1 - \gamma)[\nabla f(x + \mu v)]^2$$

$$v \leftarrow \mu v - \frac{\eta}{\sqrt{m}} \nabla f(x + \mu v)$$

$$x \leftarrow x + v \tag{11.8}$$

with μ the momentum factor, as before.

It's claimed that RMSprop is a robust classifier. We'll see below how it fared on one test. We're considering it an adaptive technique because the learning rate (η) is scaled by the square root of the running gradient mean; therefore, the effective learning rate is adjusted based on the history of the descent—it isn't fixed once and for all.

RMSprop is often used in reinforcement learning, the branch of machine learning that attempts to learn how to act. For example, playing Atari video games uses reinforcement learning. RMSprop is believed to be robust when the optimization process is *nonstationary*, meaning the statistics change in time. Conversely, a *stationary* process is one where the statistics do not change in time. Training classifiers using supervised learning is stationary, as the training set is, typically, fixed and not changing, as should be the data fed to the classifier over time, though that is harder to enforce. In reinforcement learning, time is a factor, and the statistics of the dataset might change over time; therefore, reinforcement learning might involve nonstationary optimization.

Adagrad and Adadelta

Adagrad appeared in 2011 (see "Adaptive Subgradient Methods for Online Learning and Stochastic Optimization" by John Duchi et al., *Journal of Machine Learning Research* 12[7], [2011]). At first glance, it looks quite similar to RMSprop, though there are important differences.

We can write the basic update rule for Adagrad as

$$v_i \leftarrow \frac{-\eta}{\sqrt{\sum_\tau [\nabla f(\boldsymbol{x})_i]^2}} \nabla f(\boldsymbol{x})_i$$

$$\boldsymbol{x} \leftarrow \boldsymbol{x} + \boldsymbol{v} \tag{11.9}$$

This requires some explanation.

First, notice the i subscript on the velocity update, both on the velocity, $\boldsymbol{v}$, and the gradient, $\nabla f(\boldsymbol{x})$. Here, i refers to a component of the velocity, meaning the update must be applied per component. The top of Equation 11.9 repeats for all the components of the system. For a deep neural network, this means all the weights and biases.

Next, look at the sum in the denominator of the per-component velocity update. Here, τ (tau) is a counter over *all* the gradient steps taken during the optimization process, meaning for each component of the system, Adagrad tracks the sum of the square of the gradient calculated at each step. If we're using Equation 11.9 for the 11th gradient descent step, then the sum in the denominator will have 11 terms, and so on. As before, η is a learning rate, which here is global to all components.

A variant of Adagrad is also in widespread use: *Adadelta*. (See "Adadelta: An Adaptive Learning Rate Method" by Matthew Zeiler, [2012].) Adadelta replaces the square root of the sum over all steps in the velocity update with a running average of the last few steps, much like the running average of RMSprop. Adadelta also replaces the manually selected global learning rate, η, with a running average of the previous few velocity updates. This eliminates the selection of an appropriate η but introduces a new parameter, γ, to set the window's size, as was done for RMSprop. It's likely that γ is less sensitive to the properties of the dataset than η is. Note how in the original Adadelta paper, γ is written as ρ (rho).

Adam

Kingma and Ba published *Adam*, from "adaptive moment estimation," in 2015, and it has been cited over 66,000 times as of this writing. Adam uses the square of the gradient, as RMSprop and Adagrad do, but also tracks a momentum-like term. Let's present the update equations and then walk through them:

$$m \leftarrow \beta_1 m + (1 - \beta_1) \nabla f(x)$$

$$v \leftarrow \beta_2 v + (1 - \beta_2) [\nabla f(x)]^2$$

$$\hat{m} \leftarrow \frac{m}{1 - \beta_1^t}$$

$$\hat{v} \leftarrow \frac{b}{1 - \beta_2^t}$$

$$x \leftarrow x - \frac{\eta}{\sqrt{\hat{v}} + \epsilon} \hat{m} \qquad (11.10)$$

The first two lines of Equation 11.10 define m and v as running averages of the first and second moments. The first moment is the mean; the second moment is akin to the variance, which is the second moment of the difference between a data point and the mean. Note the squaring of the gradient value in the definition of v. The running moments are weighted by two scalar parameters, β_1 and β_2.

The next two lines define $\hat{m}$ and $\hat{v}$. These are bias correction terms to make m and v better estimates of the first and second moments. Here, t, an integer starting at zero, is the timestep.

The actual step updates x by subtracting the bias-corrected first moment, $\hat{m}$, scaled by the ratio of the global learning rate, η, and the square root of the bias-corrected second moment, $\hat{v}$. The ϵ term is a constant to avoid division by zero.

Equation 11.10 has four parameters, which seems excessive, but three of them are straightforward to set and are seldom changed. The original paper suggests $\beta_1 = 0.9$, $\beta_2 = 0.999$, and $\epsilon = 10^{-8}$. Therefore, as with vanilla gradient descent, the user is left to select η. For example, Keras defaults to $\eta = 0.001$, which works well in many cases.

The Kingma and Ba paper shows via experiment that Adam generally outperforms SGD with Nesterov momentum, RMSprop, Adagrad, and Adadelta. This is likely why Adam is currently the go-to optimizer for many deep learning tasks.

Some Thoughts About Optimizers

Which optimization algorithm to use and when depends on the dataset. As mentioned, Adam is currently favored for many tasks, though properly tuned SGD can be quite effective as well, and some swear by it. While it's not possible to make a blanket statement about which is the best algorithm, for there is no such thing, we can conduct a little experiment and discuss the results.

This experiment, for which I'll present only the results, trained a small convolutional neural network on MNIST using 16,384 random samples for the training set, a minibatch of 128, and 12 epochs. The results show the mean and standard error of the mean for five runs of each optimizer: SGD, RMSprop, Adagrad, and Adam. Of interest is the accuracy of the test set and the training clock time. I trained all models on the same machine, so relative timing is what we should look at. No GPU was used.

Figure 11-11 shows the overall test set accuracy (top) and the training time (bottom) by optimizer.

On average, SGD and RMSprop were about 0.5 percent less accurate than the other optimizers, with RMSprop varying widely but never matching Adagrad or Adam. Arguably, Adam performed the best in terms of accuracy. For training time, SGD was the fastest and Adam the slowest, as we might expect, given the multiple per-step calculations Adam performs relative to the simplicity of SGD. Overall, the results support the community's intuition that Adam is a good optimizer.

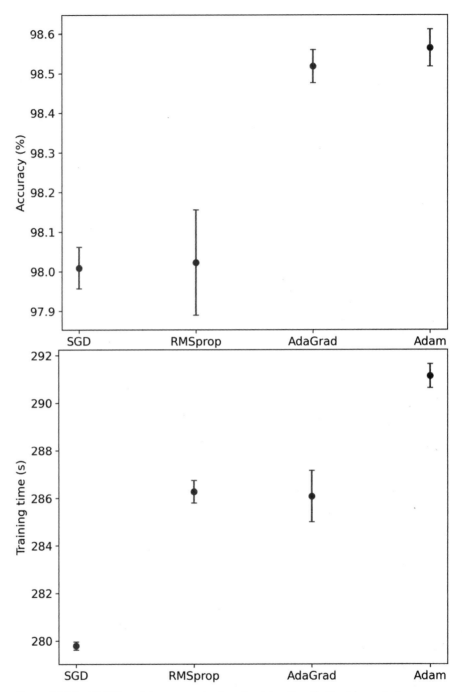

Figure 11-11: MNIST model accuracy (top) and training time (bottom) by optimizer

Summary

This chapter presented gradient descent, working through the basic form, vanilla gradient descent, with 1D and 2D examples. We followed by introducing stochastic gradient descent and justified its use in deep learning.

We discussed momentum next, both standard and Nesterov. With standard momentum, we demonstrated that it does help in training deep models (well, relatively "deep"). We showed the effect of Nesterov momentum visually using a 2D example and discussed why Nesterov momentum and stochastic gradient descent might counteract each other.

The chapter concluded with a look at the gradient descent update equations for advanced algorithms, thereby illustrating how vanilla gradient descent invites modification. A simple experiment gave us insight into how the algorithms perform and appeared to justify the deep learning community's belief in Adam's general suitability over SGD.

And, with this chapter, our exploration of the mathematics of deep learning draws to a close. All that remains is a final appendix that points you to places where you can go to learn more.

Epilogue

As the great computer scientist Edsger W. Dijkstra said, "There should be no such thing as boring mathematics." I sincerely hope you didn't find this book boring. I'd hate to offend Dijkstra's ghost. If you're still reading at this point, I suspect you did find something of merit. Good! Thanks for sticking with it. Math should never be boring.

We've covered the basics of what you need to understand and work with deep learning. Don't stop here, however: use the references in the Appendix and continue your mathematical explorations. You should never be satisfied with your knowledge base—always seek to broaden it.

If you have questions or comments, please do reach out to me at *mathfordeeplearning@gmail.com*.

GOING FURTHER

The goal of this book was to discuss the core mathematics behind deep learning, the sort of math needed to follow what deep learning is and how it operates. We've done just that in the previous 11 chapters.

In this appendix, my goal is to point you toward more. Out of necessity, we only waded in the tide pools, which are fascinating enough, but in the depths, you'll find still more beauty and elegance. What follows are pointers to help you get more out of the topics we covered.

Probability and Statistics

There are hundreds, if not thousands, of books on probability and statistics. The list here is, naturally, incomplete and not comprehensive, but it should help you expand your knowledge of these areas.

Probability and Statistics **by Michael Evans and Jeffrey Rosenthal**
A comprehensive textbook approach that's available for free here: *http://www.utstat.toronto.edu/mikevans/jeffrosenthal/book.pdf*. Evans and Rosenthal's book targets readers with exactly the sort of background that this book covers.

***Bayesian Statistics the Fun Way* by Will Kurt**   We discussed Bayes' theorem in Chapter 3. This book presents Bayesian statistics in an approachable way. Bayesian statistics is strongly related to machine learning, and you'll eventually encounter it as you progress in your studies.

***Introduction to Probability* by Joseph Blitzstein and Jessica Hwang**   Another well-liked introduction to probability, which includes Monte Carlo modeling.

***Python for Probability, Statistics, and Machine Learning* by José Unpingco**   This book provides an alternative view to the approach I took in this book. It covers slightly different topics, but still using Python and NumPy. The machine learning portion covers what I call "classic machine learning" with some mention of deep learning.

***Practical Statistics for Medical Research* by Douglas Altman**   A classic text, but still highly readable and relevant, though it comes from the era before much was done on personal computers. The focus is biostatistics, but the basics are the basics regardless of the application area.

Linear Algebra

Of all the subjects we covered, we were most unfair to linear algebra. The references here will help you appreciate the full elegance of the subject.

***Introduction to Linear Algebra* by Gilbert Strang**   A popular introductory book. It covers in greater detail many of the topics I touched on in Chapters 5 and 6.

***Linear Algebra* by David Cherney et al.**   Similar to the above, but available for free: *https://www.math.ucdavis.edu/~linear/linear-guest.pdf*.

***Linear Algebra* by Jim Hefferon**   Also available for free, and at the same level as the others: *http://joshua.smcvt.edu/linearalgebra/book.pdf*.

Calculus

We discussed calculus in Chapters 7 and 8, but we limited ourselves to differentiation only. Calculus is, of course, much more than differentiation. The other primary part of calculus is integration. We ignored integration because deep learning seldom uses it. Convolution is integration when working with continuous variables, but most integrals become summations in the digital world. The references listed here will fill in the gaps in our cursory treatment.

***Essential Calculus Skills Practice Workbook* by Chris McMullen**   This popular textbook/workbook covers differentiation and beginning integration. It includes solutions to problems. View this book as a review of Chapter 7 and an introduction to integration.

Calculus by **James Stewart**   If McMullen's book is a gentle introduction, this book is a comprehensive treatment of the topic. The book covers differentiation and integration, including multivariate calculus, that is, partial derivatives and vector calculus (see Chapter 8), and differential equations. It includes applications.

Matrix Differential Calculus with Applications in Statistics and Econometrics by **Jan Magnus and Heinz Neudecker**   Considered by some to be the standard matrix calculus reference, providing an in-depth and thorough treatment of the topic.

The Matrix Cookbook by **Kaare Brandt Petersen and Michael Syskind Pedersen**   A popular reference for matrix calculus that goes beyond what we covered in Chapter 8. You can find it here: *http://www2.imm .dtu.dk/pubdb/edoc/imm3274.pdf*.

Deep Learning

Deep learning is evolving rapidly. While some of the early, impressive "Wow, I didn't know that was even possible" results are becoming less common, the field is quietly maturing and embedding itself in almost every area of science and technology. Our world will never be the same because of what deep learning has made possible.

Deep Learning by **Ian Goodfellow, Yoshua Bengio, and Aaron Courville**   One of the first deep learning–specific textbooks and widely regarded as one of the best. It covers all the essentials but goes rather quickly at times.

A Matrix Algebra Approach to Artificial Intelligence by **Xian-Da Zhang**   A new book covering both matrices and machine learning, including deep learning. View it as a more mathematical treatment of machine learning.

Deep Learning Specialization on Coursera   Not a text, but a series of online courses with top-tier instructors. You can find the courses here: *https://www.coursera.org/specializations/deep-learning/*.

Geoffrey Hinton's Coursera lectures   In 2012, Hinton gave a lecture series on Coursera, and the lectures are well worth listening to even now. RMSprop was discussed in this series. The lectures are quite accessible and not overly math-heavy. You can find them here: *https://www.cs .toronto.edu/ hinton/coursera_lectures.html*.

Deep Learning: A Visual Approach by **Andrew Glassner**   This book covers the breadth of deep learning, from supervised learning to reinforcement learning—all without mathematics. Use it as a rapid, high-level introduction to many topics in the field.

Reddit   To keep your finger on the pulse of the deep learning community, follow the conversation on Reddit: *https://www.reddit.com/r/ MachineLearning/*.

Arxiv   Found at *https://arxiv.org/*, Arxiv is a preprint repository. The latest deep learning papers will show up here. Note, as the field progresses so rapidly, publication in peer-reviewed journals is not the norm. Rather, posting papers on arxiv.org, especially those presented at conferences, is the way to see the latest research. Arxiv is separated into categories. Those I've provided here are the ones I tend to follow, though there are others:

- Computer Vision and Pattern Recognition: *https://arxiv.org/list/ cs.CV/recent/*
- Artificial Intelligence: *https://arxiv.org/list/cs.AI/recent/*
- Neural and Evolutionary Computing: *https://arxiv.org/list/cs .NE/recent/*
- Machine Learning: *https://arxiv.org/list/stat.ML/recent/*

INDEX

eigenvector, 141
Euclidean distance, 146
event, 18
evidence (Bayesian), 59

F

F1 score, 72
false negative (FN), 251
false positive (FP), 251
Fashion-MNIST, 290
Fast Loaded Dice Roller (FLDR), 49
feature space, 105
features, 105
floating-point numbers, 5
fully connected layer, 239

G

gamma distribution, 53
Gaussian distribution, 53
geometric mean, 71
gradient
    vector field, 186
gradient descent
    Adadelta, 299
    Adagrad, 297, 299
    Adam, 297, 300
    batch training, 282
    effect of multiple minima, 281
    in 1D, 272
    in 2D, 276
    minibatch, 283, 284
    momentum, 285
        neural network, 289
    Nesterov momentum, 294
    online learning, 283
    RMSprop, 297
    stochastic, 282
    vanilla, 272

H

Hadamard product, 110
harmonic mean, 72
Hessian matrix, 211
    as Jacobian of gradient, 212
    Cholesky decomposition, 217
    critical points, 213
    optimization, 214
    quadratic approximation, 215

Hinton, Geoffrey, 297
histogram
    converting to probabilities, 43
    definition, 43
hypothesis testing, 92
    p-value, 95
    alpha, 96
    alternative hypothesis, 94
    assumptions, 95
    calculate CI, 97
    confidence interval (CI), 96
        interpreting, 97
    critical value, 97
    degrees of freedom, 95
    hypothesis, 94
    interpretation, 94
    Mann-Whitney U, 93, 99
    nonparametric, 93
    null hypothesis, 94
    one-sided, 94
    parametric, 93, 95
    statistically significant, 96
    t-test, 93
        assumptions, 95
    two-sided, 94
    warning, 96
    Welch's t-test, 95
    Wilcoxon rank sum test, 99

I

identity matrix, 132
indefinite matrix, 140
inertia, 285
inner product, 114
interquartile range, 82
inverse matrix, 138

J

joint probability, 37
    table, 33

K

k-nearest neighbors, 105
Kronecker product, 125
Kullback-Leiber divergence, 151

neural network, *continued*
  features space, 105
  feedforward, 225
  fully connected layer, 239
  hyperparameter, 283
  initialization, 42
  logistic function, 229
  minibatch, 224, 264
  momentum, 289
  pooling layer, 237
  rectified linear unit (ReLU), 226
  sigmoid function, 229
  training, 259
  weight matrix, 225
neuroevolution, 217
Newton's method, 208
  Hessian matrix, 216
  Jacobian, 209
  Taylor series approximation, 215
normal distribution, 53
not a number (NaN), 83
NumPy, 4
  array indexing, 8
  arrays on disk, 10
  broadcasting, 110
  colon, 9
  data types, 6
  defining arrays, 5
  ellipsis, 9
  matrix multiplication, 123
  special arrays, 7

O

one-hot encoding, 69
online learning, 283
optimization
  first-order, 214, 215
  neuroevolution, 217
  second-order, 214, 215
    intractable, 217
orthogonal matrix, 139
outer product, 117
outliers, 82

P

p-value, 95
principal component analysis (PCA),
    154

pooling layer
  average, 238
  information loss, 238
  maximum, 237
positive definite matrix, 140
positive semidefinite matrix, 140
posterior probability, 59
precision, 72
principal component analysis (PCA),
    154
prior probability, 60
  updating, 61
probability
  Bayes' theorem, 21, 31
    evidence, 59
    likelihood, 59
    posterior probability, 59
    prior probability, 60
    uninformed prior, 62
    updating the prior, 61
  birthday paradox, 26
  central limit theorem, 55
  chain rule, 37
  conditional, 31
  definition, 18
  distribution, 41
    Bernoulli, 48
    beta, 53, 83
    binomial, 46
    continuous, 51
    discrete, 45
    Fast Loaded Dice Roller (FLDR),
      49
    from histogram, 44
    gamma, 53
    Gaussian, 53
    lognormal, 83
    normal, 53, 83
    Poisson, 48
    probability density function, 53
    uniform, 45, 51
  enumerating the sample space, 23
  event, 18
  joint, 17, 33, 37
  joint probability table, 33
  law of large numbers, 58
  marginal, 17, 33
  mutually exclusive events, 24

Never before has the world relied so heavily on the Internet to stay connected and informed. That makes the Electronic Frontier Foundation's mission—to ensure that technology supports freedom, justice, and innovation for all people—more urgent than ever.

For over 30 years, EFF has fought for tech users through activism, in the courts, and by developing software to overcome obstacles to your privacy, security, and free expression. This dedication empowers all of us through darkness. With your help we can navigate toward a brighter digital future.

RESOURCES

Visit *https://nostarch.com/math-deep-learning/* for errata and more information.

More no-nonsense books from **NO STARCH PRESS**

PRACTICAL DEEP LEARNING

A Python-Based Introduction

BY RONALD T. KNEUSEL

464 PP., $59.95

ISBN 978-1-71850-074-7

DEEP LEARNING

A Visual Approach

BY ANDREW GLASSNER

776 PP., $99.99

ISBN 978-1-71850-072-3

NATURAL LANGUAGE PROCESSING WITH PYTHON AND SPACY

A Practical Introduction

BY YULI VASILIEV

216 PP., $39.95

ISBN 978-1-71850-052-5

BAYESIAN STATISTICS THE FUN WAY

Understanding Statistics and Probability with Star Wars, LEGO, and Rubber Ducks

BY WILL KURT

256 PP., $34.95

ISBN 978-1-59327-956-1

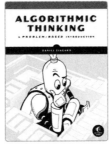

ALGORITHMIC THINKING

A Problem-Based Introduction

BY DANIEL ZINGARO

408 PP., $49.95

ISBN 978-1-71850-080-8

BEYOND THE BASIC STUFF WITH PYTHON

Best Practices for Writing Clean Code

BY AL SWEIGART

384 PP., $34.95

ISBN 978-1-59327-966-0

PHONE:
800.420.7240 OR
415.863.9900

EMAIL:
SALES@NOSTARCH.COM

WEB:
WWW.NOSTARCH.COM